POWER PLANT ENGINEERING

POWER PLANT ENGINEERING

Dr. S. Suyambazhahan
M.E., Ph.D. (IITM)
Principal, S.A. Engineering College
Avadi-Poonamallee High Road,
Veeraraghavapuram, Thiruverkadu Post
Chennai, Tamilnadu

UNIVERSITY SCIENCE PRESS
(An Imprint of Laxmi Publications Pvt. Ltd.)
An ISO 9001:2008 Company
BENGALURU • CHENNAI • COCHIN • GUWAHATI • HYDERABAD
JALANDHAR • KOLKATA • LUCKNOW • MUMBAI • RANCHI • NEW DELHI
BOSTON (USA) • NAIROBI (KENYA)

POWER PLANT ENGINEERING

Printed and bound in India
Typeset at Excellent Graphics
New Edition
ISBN 978-93-83828-52-4

Branches

Bengaluru	080-26 75 69 30	
Chennai	044-24 34 47 26,	24 35 95 07
Cochin	0484-237 70 04,	405 13 03
Guwahati	0361-254 36 69,	251 38 81
Hyderabad	040-27 55 53 83,	27 55 53 93
Jalandhar	0181-222 12 72	
Kolkata	033-22 27 43 84	
Lucknow	0522-220 99 16	
Mumbai	022-24 93 12 61	
Ranchi	0651-220 44 64	

PUBLISHED IN INDIA BY

UNIVERSITY SCIENCE PRESS
(An Imprint of Laxmi Publications Pvt.Ltd.)

An ISO 9001:2008 Certified Company
113, GOLDEN HOUSE, DARYAGANJ,
NEW DELHI - 110002, INDIA
Telephone : 91-11-4353 2500, 4353 2501
Fax : 91-11-2325 2572, 4353 2528
www.laxmipublications.com info@laxmipublications.com

C— 8969/014/07
Printed at: Repro Knowledgecast Limited, Thane

Preface

Several books on "Power Plant Engineering" are available in the market today. But the buyers must be smart enough to know which book to purchase. The first edition of the book comprising five units namely Thermal Power Plants, Hydroelectric Power Plants, Nuclear Power Plants, Gas and Diesel Power Plants, and NOn-Conventional Power Generation. It is assured that this book will satisfy the need for why it is bought and also serves as a good textbook for both students and faculties. All kinds of suggestions and constructive criticisms are earnestly solicited for enhancements. I dedicate this book to the **Almighty** for this opportunity.

This book is designed by profoundly comprehending the needs of emerging Electrical and Electronics Engineers and as per the syllabus of fourth semester course **"EE2252 Power Plant Engineering"** prescribed by Anna University, Chennai. This book is prepared in a very systematic and self-explanatory approach and the contents are presented in an elegant, comprehensive and lucid fashion with neat diagrams wherever required. The contents of the book are made available in such a way that even the beginners of the subject can understand the concepts of Power Plant Engineering and various power plants from the scrap. The content of the book is sectioned into five units namely Thermal Power Plant, Hydroelectric Power Plants, Nuclear Power Plants, Gas and Diesel Power Plants, and Non-Conventional Power Generation. Particularly topics such as boilers, turbines, condensers, pumps, Hydroelectric Power Plants nuclear Power Plants, gas and Diesel Power Plants, various non-conventional power generation techniques are elaborately dealt. All the units are provided with ample of review questions and solved problems. This book will surely satisfy the students in preparing for the University examinations and achieving prolific results and also it will quench the thirst of success in preparing other competitive examinations.

—Dr. S. Suyambazhahan

Contents

1

Thermal Power Plants

1.1 INTRODUCTION—POWER PLANT ENGINEERING

In the modern scenario, the need for electricity is increasing very rapidly. Electric power is considered as the heart of any industry. Electricity is used in our day-to-day life for lighting, heating, lifting, and cooking and so on. Therefore, it is necessary to produce electricity in large scale and also economically. The large scale power production could be achieved only by means of suitable power producing units like power plants. The most commonly used power plants are steam/thermal, Gas, Diesel, Hydroelectric and nuclear power plants. The main aspects consider while constructing or designing a power plant is the selection of proper location and appropriate equipments for the plant such that maximum output is achieved. The generated power must also be cost effective, reliable and fairly uninterrupted.

1.2 CLASSIFICATION OF POWER PLANTS

Bulk electric power is generated by special plants known as generating stations or power plants. A generating station consists of a prime mover coupled to an alternator to produce electric power.

The prime mover converts different energy forms like kinetic energy, potential energy, chemical energy, into mechanical energy. The alternator converts the mechanical energy to electrical energy.

Depending upon the energies or input converted by prime mover into mechanical energy, the power plants are classified as follows:

1. Steam power plants
2. Hydroelectric power plants
3. Diesel power plants
4. Nuclear power plants.

1.3 SOURCES OF ENERGY

The energy sources can be classified as:

- Conventional energy sources or Non-renewable energy sources
- Non-conventional energy sources or Renewable energy sources

1.3.1 Conventional Sources of Energy

The conventional resources are finite and exhaustible. Once consumed, these sources cannot be replaced by others. Examples are coal, timber, petroleum, lignite, natural gas, fossil fuels, nuclear fuels etc.

The examples for conventional sources of energy are:

(*i*) Fossil fuel energy

(*ii*) Nuclear energy

(*iii*) Hydro energy

Fossil Fuel Energy

Fossil fuel is an invaluable source of energy produced due to chemical changes taking place in the absence of oxygen in plants and animals that have been buried deep in the earth's crust for many million years. Fossil fuels like coal, petroleum and natural gas are formed in this manner. These are conventional sources of energy. For example, energy from, Petroleum, natural gas, coal, nuclear energy, etc.

Nuclear Energy

Nuclear energy is obtained by fission of nuclear material. Uranium is a fairly common element in earth's crust. Naturally found uranium contains 0.72% of uranium 235 (U-235) by mass rest being Uranium 238 (U-238) and small amount of isotopes. Isotopes are atoms of same element having different atomic weights. Plutonium and thorium are another alternate fuel for creating nuclear energy.

Hydro Energy

It is the energy possessed by the water. The potential energy of water stored in a reservoir at high level is converted into high kinetic energy water by passing to the turbine through the nozzles to produce mechanical energy. Then the mechanical energy is converted into electrical energy by coupling turbine with generator. Thus the hydro energy is obtained for producing electricity. The conventional power plants are producing electricity using conventional energy

1.3.2 Non-Conventional Sources of Energy

The non-conventional sources are being continuously produced in nature and are not exhaustible. Examples are wood, geothermal energy, wind energy, tidal energy, nuclear fusion, gobar gas, biomass, solar energy etc.

The examples for non-conventional sources of energy are:

(*i*) Solar energy

(*ii*) Wind energy

(*iii*) Geothermal energy

(*iv*) Ocean energy such as tidal energy, wave energy

(*v*) Biomass energy such as gobar gas.

It is evident that all energy resources based on fossil fuels has limitations in availability and will soon exhaust. Hence the long-term option for energy supply lies only with non-conventional energy sources. The non-conventional power plants are producing electricity using non-conventional energy.

Renewable Energy

Renewable energy is energy which is generated from natural sources *i.e.,* sun, wind, rain, tides and can be generated again and again as and when required. They are available in plenty and by far most the cleanest sources of energy available on this planet. For example, Energy that we receive from the sun can be used to generate electricity. Similarly, energy from wind, geothermal, biomass from plants, tides can be used this form of energy to another form.

Advantages

- The sun, wind, geothermal, ocean energy are available in the abundant quantity and free to use.
- The non-renewable sources of energy that we are using are limited and are bound to expire one day.
- Renewable sources have low carbon emissions, therefore they are considered as green and environment friendly.
- Renewable helps in stimulating the economy and creating job opportunities. The money that is used to build these plants can provide jobs to thousands to lakhs of people.
- You don't have to rely on any third country for the supply of renewable sources as in case of non-renewable sources.
- Renewable sources can cost less than consuming the local electrical supply. In the long-run, the prices of electricity are expected to soar since they are based on the prices of crude oil, so renewable sources can cut your electricity bills.
- Various tax incentives in the form of tax waivers, credit deductions are available for individuals and businesses who want to go green.

Disadvantages

- It is not easy to set up a plant as the initial costs are quite steep.
- Solar energy can be used during the daytime and not during night or rainy season.
- Geothermal energy which can be used to generate electricity has side effects too. It can bring toxic chemicals beneath the earth surface onto the top and can create environmental changes.
- Hydroelectric provide pure form of energy but building dams across the river which is quite expensive can affect natural flow and affect wildlife.
- To use wind energy, you have to rely on strong winds therefore you have to choose suitable site to operate them. Also, they can affect bird population as they are quite high.

Non-Renewable Energy

Renewable energy is energy which is taken from the sources that are available on the earth in limited quantity and will vanish fifty–sixty years from now. Non renewable sources are not environmental friendly and can have serious affect on our health. They are called non-renewable because they can be re-generated within a short span of time. Non-renewable sources exist in the form of fossil fuels, natural gas, oil and coal.

Advantages

- Non-renewable sources are cheap and easy to use. It can easily fill up in car tank and power the motor vehicle.
- It can use small amount of nuclear energy to produce large amount of power.
- Non-renewable have little or no competition at all. For example, in a battery driven car, battery gets discharged then it won't be able to charge in the middle of the road rather it is easy to find a gas pumping station.
- It is very cheap when converting from one type of energy to another.

Disadvantages

- Non-renewable sources will expire some day and we have to us our endangered resources to create more non-renewable sources of energy.

- The speed at which such resources are being utilized can have serious environmental changes.
- Non-renewable sources release toxic gases in the air when burnt which are the major cause for global warming.
- Since these sources are going to expire soon, prices of these sources are soaring day by day.

1.3.3 Comparison of Sources of Energy

Sl. No.	*Renewable energy sources*	*Non-renewable energy sources*
1.	They don't cause pollution to atmosphere.	They cause pollution to atmosphere.
2.	Their efficiency is comparatively less.	Their efficiency is comparatively high.
3.	They are available at low intensities.	They are available at high intensities.
4.	Their initial cost is comparatively less.	Their initial cost is comparatively high.
5.	Their running cost is less.	Their running cost is more.

1.4 HISTORY OF THERMODYNAMICS

Thermodynamics is a branch of physics and of chemistry that studies the effects of changes in temperature, pressure, and volume on physical systems at the macroscopic scale by analyzing the collective motion of their particles using statistics. Roughly, heat means "energy in transit" and dynamics relates to "movement"; thus, in essence thermodynamics studies the movement of energy and how energy instills movement. Historically, thermodynamics developed out of need to increase the efficiency of early steam engines.

A brief history of thermodynamics begins with Otto von Guericke who in 1650 built and designed the world's first vacuum pump and created the world's first ever vacuum (known as the Magdeburg hemispheres). He was driven to make a vacuum in order to disprove Aristotle's long-held supposition that 'nature abhors a vacuum'. Shortly thereafter, Irish physicist and chemist Robert Boyle had learned of Guericke's designs and in 1656, in coordination with English scientist Robert Hooke, built an air pump. Using this pump, Boyle and Hooke noticed a correlation between pressure, temperature, and volume. In time, Boyle's Law was formulated, which states that pressure and volume are inversely proportional. Then, in 1679, based on these concepts, an associate of Boyle's named Denis Papin built a bone digester, which was a closed vessel with a tightly fitting lid that confined steam until a high pressure was generated.

Later designs implemented a steam release valve that kept the machine from exploding. By watching the valve rhythmically move up and down, Papin conceived of the idea of a piston and a cylinder engine. He did not, however, follow through with his design. Nevertheless, in 1697, based on Papin's designs, engineer Thomas Savery built the first engine. Although these early engines were crude and inefficient, they attracted the attention of the leading scientists of the time. One such scientist was Sadi Carnot, the "father of thermodynamics", who in 1824 published *Reflections on the Motive Power of Fire*, a discourse on heat, power, and engine efficiency. The paper outlined the basic energetic relations between the Carnot engine, the Carnot cycle, and Motive power. This marks the start of thermodynamics as a modern science.

1.4.1 Temperature and Heat

Although they are closely related, heat (denoted ΔQ) and temperature (denoted T) are NOT the same. Temperature is a measure of the kinetic energy of vibration of the molecules that make up a substance. Because the molecules are connected to one another in vastly different ways from one substance to the next, the simplest and most useful way to think of temperature is based upon a comparison system. The idea is the following. If two substances are at the same temperature, then nothing will happen when they are placed in thermal contact with one another. That is, neither will heat up of cool down. If they are at different temperatures, then the hotter one will cool down and the cooler one will heat up ... until they reach the same temperature! We say the two objects are in thermodynamic equilibrium. (Doesn't that sound more impressive than the same temperature?) This simple idea is encoded in the *Zeroth Law of Thermodynamics:*

If object A is in thermodynamic equilibrium with object B, and object B is in thermodynamic equilibrium with object C, then object A is in thermodynamic equilibrium with object C.

You can think of object *B* as a thermometer. The two most commonly used temperature scales are the Fahrenheit scale and the Celsius scale. On the Fahrenheit scale, water freezes at 32 °F and boils at 212 °F. On the Celsius scale (also referred to as the centigrade scale) water freezes at 0°C and boils at 100°C. To convert from one scale to the other, use

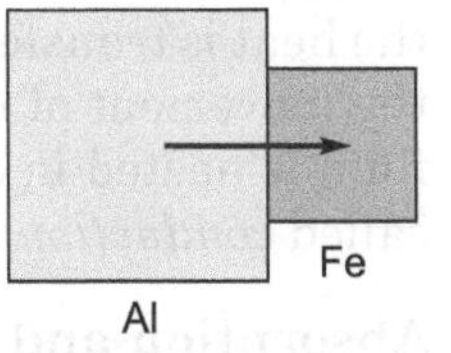

$$T^\circ_C = 5/9\,(T^\circ_F - 32)$$

For a quick estimate of the Celsius temperature, subtract 30 from the Fahrenheit temperature and divide by two. For the reverse conversion estimate, multiply the Celsius temperature by 2 and add 30. There is a third scale used in science that is important. It is the Kelvin or Absolute scale. The Kelvin temeprature is found from the Celsius temperature by simply adding 273.

$$T^\circ_K = T^\circ_C + 273$$

On the Kelvin scale, water freezes at the scorching temperature of 273K. The coldest temperature possible is 0°K, often referred to as absolute zero. (Absolute zero is the temperature at which all atomic and molecular motion ceases, in the "classical theory". The more accurate "quantum theory" paints a more complicated picture).

Heat

Heat is one of the many forms of energy. We do not usually refer to the absolute amount of heat energy in an object. That can be complicated and is generally not a useful quantity. Instead, we refer to the amount of heat that "flows into" or "flows out of" an object. Let's consider an experiment where we place a hot block of aluminum (Al) in contact with a cold block of iron (Fe). For now, assume the blocks have exactly the same mass. (So why are they different sizes in the picture?) Heat *flows out of* the hot aluminum (and it's temperature decreases) and *flows into* the cold iron (and it's temperature increases). Of course, heat could also flow into and out of the surrounding environment. But we eliminate this by carefully *insulating* the system. This process continues until the block reach the same temperature. Since the masses are the same, is the final temperature exactly halfway between the original temperatures? NO! How much the temperature increases or decreases depends upon the amount (specifically the mass) of the substance AND upon the substance itself. It takes more energy to heat up or cool off the block of Al than it does for Fe. We say the Al has a greater heat capacity. For a substance of mass m, the amount of heat energy ΔQ required to raise it's temperature by an amount ΔT, is given by:

$$\Delta Q = m\,c\,\Delta T$$

Conduction

Conduction is the heat transfer process in which heat "flows" from a hot region to a cooler one through direct physical contact. Consider a rectangular block of thickness Δx that has one face at temperature T_1 and the opposite face at a lower temperature T_2. Heat flows from the hot face to the cooler one. But what exactly is flowing? Early in the study of heat, it was thought that a substance called caloric flowed. What is actually happening is that the vibrational energy (heat energy) of the molecules are being transferred from the hotter regions (faster vibrations) to the cooler regions (slower vibrations). This completely natural process is what causes heat to flow from hot to cold. (Although energy is conserved, this is more than just conservation of energy. If you are interested in the First and Second Laws of Thermodynamics and how they are related to the fascinating topic of Entropy.

Convection

How does heat "flow" from one place to another? It does so through some combination of three basic processes. One method you studied in SCI 100 and is called *convection.* In this process, the heat is transferred via a flowing of a fluid. The earth's flowing magma, blowing winds, and the movement of radiator fluid in your car's engine are all examples of this process. But, the fluid is heated in one area and cooled in another through direct physical contact in a process called *conduction.* The third process is *radiation* and is discussed below.

Absorption and Radiation

All objects above a temperature of absolute zero radiate electromagnetic energy. At room temperature, almost all of this energy is in the infra-red (IR) region and is perceived as heat. The thermal vibrations of the electrons in the substance create the electromagnetic radiation. As objects become hotter, the frequency at which most of the radiation is emitted increases. The heating elements in an electric stove begin to glow dull red and eventually become orange. If they became even hotter, they would glow yellow and eventually white.

Objects also absorb radiation energy. When radiation strikes an object, one can think of the alternating electric field in the electromagnetic wave as shaking the electrons back and forth, creating thermal or heat energy. It is just the reverse of the radiation process. Objects radiate energy as well as they absorb it. The rate at which objects radiate depends upon several factors. It depends upon the surface area (A), the temperature of the object (T_{obj}), and upon the *emissivity,* ε, of the material. The emissivity is a number between 0 and 1. Very reflective materials (typically white) have 'ε' close to zero, while very absorptive surfaces (typically black) have 'ε' close to one. The rate of absorption depends upon these same factors, but it is the temperature of the surrounding environment (T_{env}) that conributes to absorption. For an object at a Kelvin temperature T_{obj}, the rate of radiation (measured in watts) is:

$$\Delta Q / \Delta t = \varepsilon \, \sigma \, A \, T_{obj}^{4} \text{ (heat radiated away)}$$

where A is the surface area and $\sigma = 5.67 \times 10^{-8}$ W / m^2 °K^4 is Stefan's constant. The exact same equation applies to absorption when the object is surrounded by an environment of temperature T_{env}. The *net heat* flowing into the object is the difference between radiation and absorption. So it is:

$$(\Delta Q / \Delta t)_{\text{net}} = \varepsilon \, \sigma \, A \, (T_{obj}^{4} - T_{env}^{4}) \text{ (net heat)}$$

Note that when the object is at the same temperature as its surroundings, there is no net heat flow and the object stays at the same temperature ... good common sense.

Here the study particularly interested in the absorption of heat from the sun. As it is pointed out in the previous module, the sun provides an intensity of about 1400 W/m^2 just outside the earth's atmosphere. How much of that reaches the earth is a function of such variables as cloud cover, humidity and where the sun is in the sky.

1.4.2 Laws of Thermodynamics

The laws of thermodynamics describe some of the fundamental truths of thermodynamics observed in our Universe. In simplest terms, the laws of Thermodynamics dictate the specifics for the movement of heat and work.

Basically, the First Law of Thermodynamics is a statement of the conservation of energy – the Second Law is a statement about the direction of that conservation – and the Third Law is a statement about reaching Absolute Zero (0°K).

- Zeroth law refers to thermodynamic equilibrium and temperature.
- First law refers to heat, work and energy.
- Second law refers to entropy.
- Third law of thermodynamics refers to absolute temperature.

The first law of thermo dynamics relates the various forms of kinetic and potential energy in a system to the work which a system can perform and to the transfer of heat. This law is sometimes taken as the definition of internal energy and introduces an additional state variable, enthalpy.

The second law of thermodynamics introduces state variable called entropy. The second law stipulates that the total entropy of a system plus its environment cannot decrease; it can remain constant for a reversible process but must always increase for an irreversible process.

First Law of Thermodynamics

"The change in a system's internal energy is equal to the difference between heat added to the system from its surroundings and work done by the system on its surroundings."

Physicists typically use uniform conventions for representing the quantities in the first law of the thermodynamics. They are:

- U_1 (or U_i) = initial internal energy at the start of the process.
- U_2 (or U_f) = final internal energy at the end of the process.
- $\Delta U = U_2 - U_1$ = change in internal energy (used in cases where the specifies of beginning and ending internal energies are irrelevant)
- Q = heat transferred into ($Q > 0$) or out of ($Q < 0$) the system.
- W = work performed by the system ($W > 0$) or on the system ($W < 0$).

This yields a mathematical representation of the first law which proves very useful and can be rewritten in a couple of useful ways:

$$U_2 - U_1 = \Delta U = Q - W$$
$$Q = \Delta U + W$$

The analysis of a thermodynamic process. Generally involves analyzing a situation where one of these quantities is either 0 or at least controllable in a reasonable manner. For example, in an adiabatic process, the heat transfer (Q) is equal to 0 while in an isochoric process the work (W) is equal to 0.

Law of Conservation of Energy

The first law of thermodynamics is often called the Law of Conservation of Energy. This law suggests that energy can be transferred from one system to another in many forms. Also, it cannot be created or destroyed.

Thus, the total amount of energy available in the Universe is constant. Einstein's famous equation (written below) describes the relationship between energy and matter:

$$E = mc^2$$

In the equation above, energy (E) is equal to matter (m) times the square of a constant (c). Einstein suggested that energy and matter are interchangeable. His equation also suggests that the quantity of energy and matter in the Universe is fixed.

Second Law of Thermodynamics

This principle was first illuminated by the French physicist and engineer Sadi Carnot. as he developed his Carnot cycle engine in 1824, and was later formalized as a law of thermodynamics by German physicist Rudolf Clausius. It is impossible for a process to have as its sole result the transfer of heat from a cooler body to a hotter one.

Heat cannot be transfer from a colder to a hotter body. As a result of this fact of thermodynamics, natural processes that involve energy transfer must have one direction, and all natural processes are irreversible.

This law also predicts that the entropy of an isolated system always increases with time. Entropy is the measure of the disorder or randomness of energy and matter in a system. Because of the second law of thermodynamics both energy and matter in the Universe are becoming less useful as time goes on.

In practical applications, this law means that any heat engine or similar device based upon the principles of thermodynamics cannot. Even in theory, be 100% efficient.

Entropy and the Second Law of Thermodynamics

The second law of thermodynamics is perhaps the most popular outside of the realm of physics, because it is closely related to the concept of entropy, or the disorder created during a thermodynamic process.

Reformulated as a statement regarding entropy. the second law reads:

Mathematical Representation of the First Law

In any closed system, the entropy of the system will either remain constant or increase.

In other words, each time a system goes through a thermodynamic process, the system can never completely return to precisely the same state it was in before. This is one definition used for the arrow of time, since entropy of the universe will always increase over time according to the second law of thermodynamics.

Entropy is the quantitative measure of disorder in a system. The concept comes out of thermodynamics, which deals with the transfer of heat energy within a system. Instead of talking about some form of "absolute entropy", physicists generally talk about the change in entropy that takes place in a specific thermodynamic process.

Calculating Entropy

In an isothermal process, the change in entropy (delta S) is the change in heat (Q) divided by the absolute temperature (T):

$$\Delta S = dQ / T$$

In any reversible thermodynamic process, it can be represented in calculus as the integral form a processes initial state to final state of dQ/T.

The SI units of entropy are J/K (joules/degrees Kelvin).

Entropy and the Second Law of Thermodynamics

One way of stating the second law of thermodynamics is: In any closed system. the entropy of the system will either remain constant or increase. One way to view this is that adding heat to a system causes the molecules and atoms to speed up.

It may be possible (though tricky) to reverse the process in a closed system (*i.e.*, without drawing any energy from or releasing energy somewhere else) to reach the initial state, but you can never get the entire system "less energetic" than it started ... the energy just doesn't have any place to go.

Misconceptions About Entropy

This view of the second law of thermodynamics is very popular, and it has been misused. Some argue that the second law of thermodynamics means that a system can never become more orderly. Not true.

It just means that in order to become more orderly (for entropy to decrease), you must transfer energy from somewhere outside the system, such as when a pregnant woman draws energy from food to cause the fertilized egg to become a complete baby, completely in line with the second line's provisions.

Other Second Law Formulations

1. **Kelvin Plank Statement:** A cyclic transformation whose only final result is to transform heat extracted from a source which is at the same temperature throughout into work is impossible.

2. **Clausius Statement:** A cyclic transformation whose only final result is to transfer heat from a body at a given temperature to a body at a higher temperature is impossible.

Third Law of Thermodynamics

The third law of thermodynamics is essentially a statement about the ability to create an absolute temperature scale, for which absolute zero is the point at which the internal energy of a solid is precisely 0.

The third law of thermodynamics states that if all the thermal motion of molecules (kinetic energy) could be removed. a state called absolute zero would occur. Absolute zero results in a temperature of 0 Kelvin or – 273.15° Celsius.

Absolute Zero = 0 Kelvin = – 273.15° Celsius

Various sources show the following three potential formulations of the third law of thermodynamics:

1. It is impossible to reduce any system to absolute zero in a finite series of operations.
2. The entropy of a perfect crystal of an element in its most stable form tends to zero as the temperature approaches absolute zero.
3. As temperature approaches absolute zero, the entropy of a system approaches a constant.

The Universe will attain absolute zero when all energy and matter is randomly distributed across space. The current temperature of empty space in the Universe is about 2.7 Kelvins.

1.4.3 Thermodynamic Processes

The change of state is called thermodynamics. For example, take 1 kg of air in a system, it's initial state, Pressure P_1, Volume V_1 temperature T_1. If the air is allowed to expand or compress, the Pressure, Volume and Temperature will be changed to a final state of P_2, V_2 and T_2.

The thermodynamic process are broadly classified into:

1. Non-flow process
2. Flow process

Examples for Non-flow process are:

1. Constant volume process or isochoric process

2. Constant pressure process or isobaric process
3. Constant temperature process or isothermal process
4. Adiabatic process or isentropic process
5. Polytrophic process etc.,

Throttling process is one of the flow processes (*e.g.* Flow through nozzles). Thermodynamic cycles can be obtained by combining more than one process. Also if the initial state is equal to final state then the cycle is called closed cycle otherwise open cycle.

1.4.4 Thermodynamic Cycles

The laws of thermodynamics are fundamental truths based on the study of energy exchanges between a system and its surroundings. The first law of thermodynamics states that energy can only be converted from one state to another, but cannot be created or destroyed. Energy referred to here is that of work, heat and internal energy. Work is done when a force acts through a distance and the transfer of energy from the system to its surroundings results. Heat is an energy transfer which takes place when two regions of different temperatures are connected by a thermal conductor. Internal energy refers to the energy stored in a body, and is therefore equal to the heat flow into the system minus the work done by the system on its surroundings.

"The second law of thermodynamics states that heat, on its own accord, flows from regions of high temperatures to regions of low temperatures." When applying the law to heat engines, it can be stated that no heat engine can be 100% efficient. [Perpetual Motion Machine - PMM]

When looking at the two laws together, it is noted that energy is being constantly degraded; in a thermodynamic process there is always less energy available for doing work, not more. This leads to the definition of entropy which always increases as energy becomes less available for doing work.

A thermodynamic cycle is a series of thermodynamic processes transferring heat and work, while varying pressure, temperature, and other state variables, eventually returning a system to its initial state.

Two primary classes of thermodynamic cycles are:

1. Power cycles and
2. Heat pump cycles

Power cycles are cycles which convert some heat input into a mechanical work output, while heat pump cycles transfer heat from low to high temperatures using mechanical work input.

Cycles composed entirely of quasistatic processes can operate as power or heat pump cycles by controlling the process direction. On a pressure-volume or Temperature—Entropy (T-S) diagram, the clockwise and counterclockwise directions indicate power and heat pump cycles, respectively.

Thermodynamic power cycles are the basis for the operation of heat engines. Which supply most of the world's electric power and run almost all motor vehicles.

Power cycles can be divided according to the type of heat engine they seek to model.

The most common cycles that model internal combustion engines are the:

1. Otto cycle, which models gasoline engines
2. Diesel cycle, which models diesel engines
3. Brayton cycle, which models gas turbines
4. Rankine cycle, which models steam turbines

The first law of thermodynamics dictates that the net heat input is equal to the network output over any cycle. The repeating nature of the process path allows for continuous operation, making the cycle an important concept in thermodynamics. Thermodynamic cycles often use quasistatic processes to model the workings of actual devices.

A thermodynamic cycle is a closed loop on a *P-V* diagram. A *P-V* diagram's *Y*-axis shows pressure (*P*) and *X*-axis shows volume (*V*). (Figure 1.1)

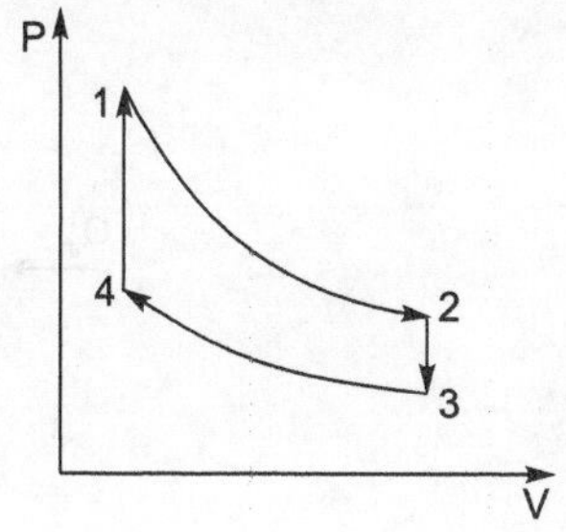

Fig. 1.1. *P-V* diagram of an Otto cycle

The area enclosed by the loop is the work (*W*) done by the process:

$$W = P\,dV$$

This work is equal to the balance of heat (*Q*) transferred into the system:

$$W = Q = Q_{in} - Q_{out}$$

The equation makes a cyclic process similar to an isothermal process: even though the internal energy changes during the course of the cyclic process, when the cyclic process finishes the system's energy is the same as the energy it had when the process began.

If the cyclic process moves clockwise around the loop, then it represents a heat engine, and *W* will be positive. If it moves counterclockwise then it represents a heat pump, and *W* will be negative.

A thermodynamic cycle which is applied typically to a steam power plant, as shown in the illustration, the cycle has four phases. (Figure 1.2)

1. Heat addition bcde in a boiler at constant pressure *PI* changing water at *b* to superheated steam at *e*.
2. Isentropic expansion *ef* in a prime mover from initial pressure P_1 to backpressure P_2.
3. Heat rejection *fa* in a condenser at constant pressure P_2 with wet steam at *f* converted to saturated liquid at *a*, and
4. Isentropic compression ab of water in a feed pump from pressure P_2 to pressure P_1.

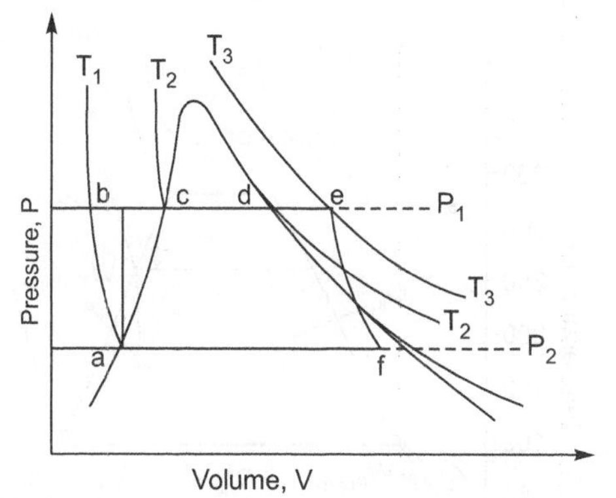

Fig. 1.2. *P-V* diagram of a steam power plant

This cycle more closely approximates the operations in a real steam power plant than does the Carnot cycle.

Rankine Cycle

The Rankine cycle is a thermodynamic cycle used to generate electricity in many power stations, and is the real-world approach to the Carnot cycle. Wet Steam is produced in boiler then the wet steam is converted into superheated steam in a super heater, and then expanded in a steam turbine. The steam turbine drives a generator, to convert the work into electricity. (Figure 1.3).

Power plants generate electrical power by using fuels like coal, oil or natural gas. A simple power plant consists of a boiler, turbine, condenser and a pump.

Fuel, burned in the boiler and super heater, heats the water to generate steam. The steam is then heated to a superheated state in the super heater.

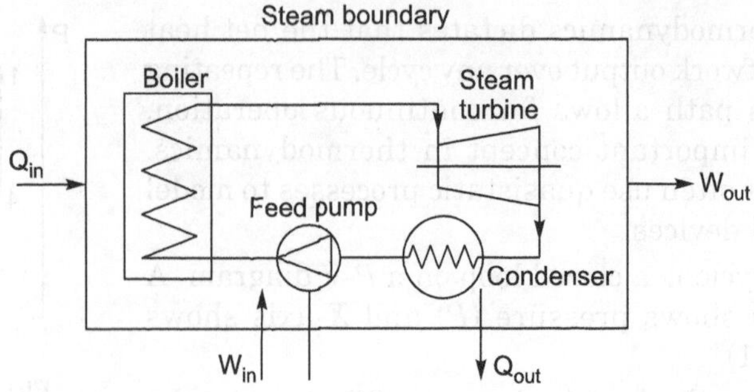

Fig. 1.3. Steam turbine power plant

This steam is used to rotate the turbine which powers the generator. Electrical energy is generated when the generator windings rotate in a strong magnetic field. After the steam leaves the turbine it is cooled to its liquid state in the condenser. The liquid is pressurized by the pump prior to going back to the boiler A simple power plant is described by a Rankine Cycle.

Processes of the Rankine Cycle

Rankine cycle is a heat engine with vapour power cycle. The common working fluid is water. There are four processes in the Rankine cycle; these states are identified by number in the diagram to the right. (Figure 1.4).

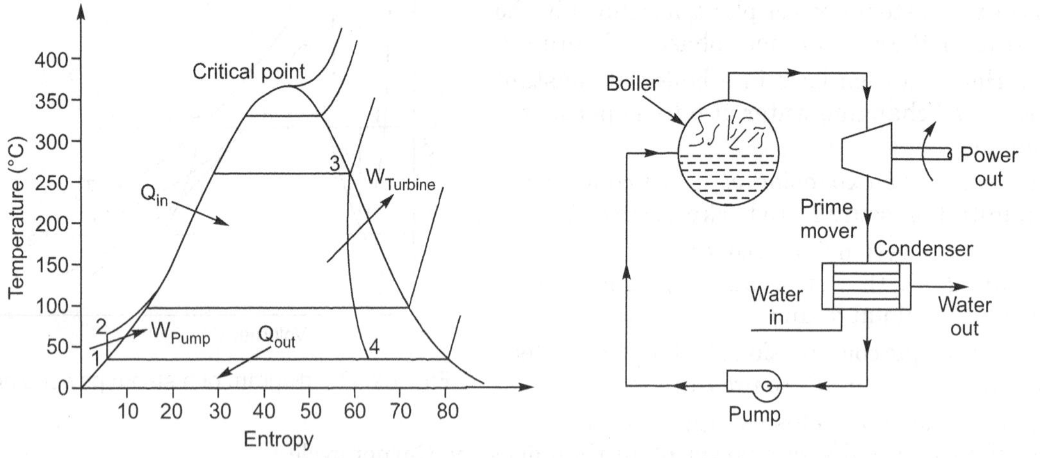

Fig. 1.4. Rankine cycle on *T-S* diagram

(*i*) *Process* 1–2: The working fluid is pumped from low to high pressure. as the fluid is a liquid at this stage the pump requires little input energy.

(*ii*) *Process* 2–3: The high pressure liquid enters a boiler where it is heated at constant pressure by an external heat source to become a dry saturated vapour.

(*iii*) *Process* 3–4: The dry saturated vapour expands through a turbine. generating power. This decreases the temperature and pressure of the vapour, and some condensation may occur.

(*iv*) *Process* 4–1: The wet vapour then enters a condenser where it is condensed at a constant pressure and temperature to become a saturated liquid. The pressure and temperature of the condenser is fixed by the temperature of the cooling coils as the fluid is undergoing a phase change.

The TS diagram of a typical Rankine cycle operating between pressures of 0.06 bar and 50 bar.

In an ideal Rankine cycle the pump and turbine would be isentropic. *i.e.*, the pump and turbine would generate no entropy and hence maximize the network output. Processes 1–2 and 3–4 would be represented by vertical lines on the *T-S* diagram and more closely resemble that of the Carnot cycle. The Rankine cycle shown here prevents the vapour ending up in the superheat region after the expansion in the turbine, which reduces the energy removed by the condensers.

Work output of the cycle (Steam turbine). W_1 and work input to the cycle (Pump). W_2 are:

$$W_1 = m(h_1 - h_2)$$

$$W_2 = m(h_3 - h_4)$$

where m is the mass flow of the cycle. Heat supplied to the cycle (boiler). Q_1 and heat rejected from the cycle (condenser). Q_2 are:

$$Q_1 = m(h_1 - h_4)$$

$$Q_2 = m(h_2 - h_3)$$

The network output of the cycle is:

$$W = W_1 - W_2$$

The thermal efficiency of a Rankine cycle is:

$$\eta = W/Q_1$$

The efficiency of the Rankine cycle is not as high as Carnot cycle but the cycle has less practical difficulties and more economic.

Regeneration helps improve the Rankine cycle efficiency by preheating the feedwater into the boiler. Regeneration can be achieved by open feedwater heaters or closed feedwater heaters.

In open feedwater heaters, a fraction of the steam exiting a high pressure turbine is mixed with the feedwater at the same pressure. In closed system, the steam bled from the turbine is not directly mixed with the feedwater, and therefore, the two streams can be at different pressures.

Organic Rankine Cycle

"Organic" is a term used in chemistry to describe a class of chemicals that includes Freon and most of the other common refrigerants. A Rankine cycle is a closed circuit steam cycle. An "organic" Rankine cycle uses a heated chemical instead of steam as found in the Rankine Cycle. Chemicals used in the Organic Rankine Cycle include Freon, butane, propane, ammonia and the new environmentally-friendly" refrigerants.

Carnot Cycle

The Carnot Cycle has been described as being the most efficient thermal cycle possible, wherein there is no heat losses, and consisting of four reversible processes, two isothermal and two adiabatic. It has also been described as a cycle of expansion and compression of a reversible heat engine that does works with no loss of heat.

By using the second law of thermodynamics it is possible to show that no heat engine can be more efficient than a reversible heat engine working between two fixed temperature limits. This heat engine is known as Carnot cycle and consists of the following processes: (Figure 1.5) (*i*) 1 to 2 : Isentropic expansion

(*ii*) 2 to 3 : Isothermal heat rejection

(*iii*) 3 to 4 : Isentropic compression

(*iv*) 4 to 1 : Isothermal heat supply

The supplied heat to the cycle per unit mass flow is $Q_1 = T_1\Delta S$

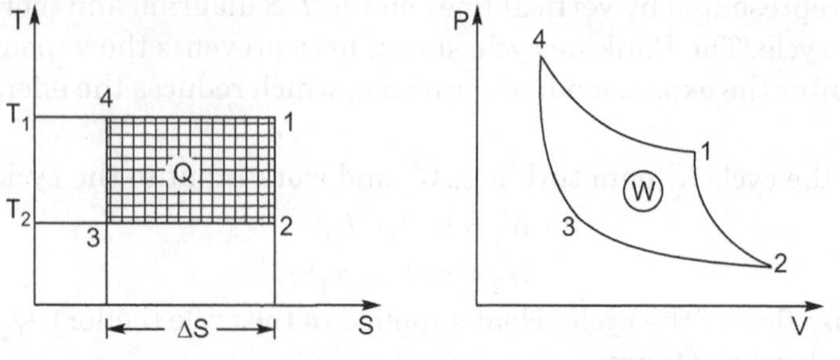

Fig. 1.5. Carnot cycle *T-S* and *P-V* diagram

The rejected heat from the cycle per unit mass flow is $Q_2 = T_2\Delta S$

By applying the first law of thermodynamics to the cycle, we obtain

$$Q_1 - Q_2 = W = 0$$

And the thermal efficiency of the cycle will be:

$$\eta = W/Q_1 = 1 - \frac{T_2}{T_1}$$

Due to mechanical friction and other irreversibility no cycle can achieve this efficiency.

The gross work output of cycle, *i.e.*, the work done by the system is:

$$W_g = W(4 - 1) + W(1 - 2)$$

and work ratio is defined as the ratio of the network, W to the gross work output. W_g.

i.e., $$W/W_g$$

The Carnot cycle has a low work ratio. Although this cycle is the most efficient system for power generation theoretically, it cannot be used in practice. There are several reasons such as low work ratio, economical aspects and practical difficulties.

Brayton Cycle

The Brayton cycle is a thermodynamic cycle that describes the workings of the gas turbine engine, basis of the jet engine and others. It is named after George Brayton (1830–1892), the American engineer who developed it, although it was originally proposed and patented by Englishman John Barber in 1791. It is also sometimes known as the Joule cycle.

A Brayton-type engine consists of three components:

1. A gas compressor
2. A mixing chamber
3. An expander

The Ideal Brayton cycle has:

- *isentropic process*—Ambient air is drawn into the compressor, where it is pressurized.
- *isobaric process*—The compressed air then runs through a combustion chamber, where fuel is burned, heating that air-a constant-pressure process. since the chamber is open to flow in and out.

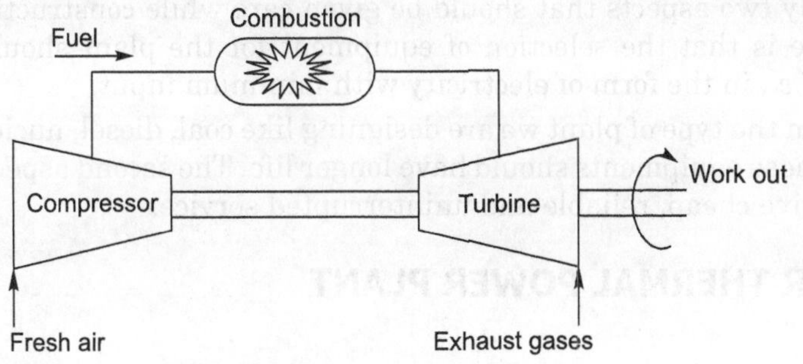

Fig. 1.6. Layout of thermal power plant

- *isentropic process*—The heated, pressurized air then gives up its energy, expanding through a turbine (or series of turbines). Some of the work extracted by the turbine is used to drive the compressor.
- *isobaric process*—Heat Rejection (in the atmosphere).

The Actual Brayton Cycle has

- adiabatic process—Compression.
- isobaric process—Heat Addition.
- adiabatic process—Expansion.
- isobaric process—Heat Rejection.

The Brayton cycle depicts the air-standard model of a gas turbine power cycle. (Figure 1.7) The four steps of the cycle are:

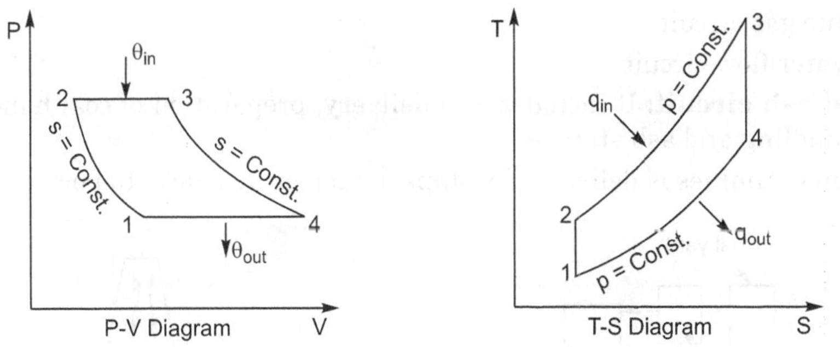

Fig. 1.7. Idealized Brayton *P-V* and *T-S* diagram

- (1–2) Isentropic Compression
- (2–3) Reversible Constant Pressure Heat Addition
- (3–4) Isentropic Expansion
- (4–1) Reversible Constant Pressure Heat Rejection

Generating Power Plant

In the modem world the need of electricity is increasing very rapidly from domestic purposes to large industrial sector so it became necessary to produce electricity in large scale and economically.

This large scale energy production could be achieved by means of suitable power producing units named 'Power Plants or Electric Power Generating Stations or Electric Power Plants.

There are mainly two aspects that should be given care while constructing or designing a plant, the first one is that the selection of equipment for the plant should be able to give maximum output *i.e.*, in the form of electricity with minimum input.

Input depends on the type of plant we are designing like coal, diesel, nuclear power etc. and at the same time these equipments should have longer life. The second aspect is that the plant should be able to give cheap, reliable and uninterrupted service.

1.5 STEAM OR THERMAL POWER PLANT

Description

The layout of steam/thermal power plant is shown in Figure 1.8. The fuel used in the steam power plant is coal.

Coal contains elements which get oxidized during reaction with oxygen, supplied by air in the furnace of the steam generator, with the release of a large amount of energy.

The large amount of energy released during the reaction is transferred to water which is converted to high pressure steam. This high pressure steam is made to undergo expansion to low pressure in a turbine, thereby effecting a conversion of low grade heat energy into high grade mechanical work which is manifested as "Torque" at the turbine shaft.

This torque is transferred directly to the rotor of the electrical generator. Electrical energy is thus produced. The potential of the electrical energy is then raised by the transformers.

The major components and their functions can be explained in four circuits.

They are:

- Fuel (coal) and ash circuit
- Feedwater and steam flow circuit
- Air and flue gas circuit
- Cooling water flow circuit

1. Fuel and ash circuit: It includes coal delivery, preparation of coal handling to boiler furnace, ash handling and ash storage.

The coal from coalmines is delivered by ships, trains or by trucks to power station.

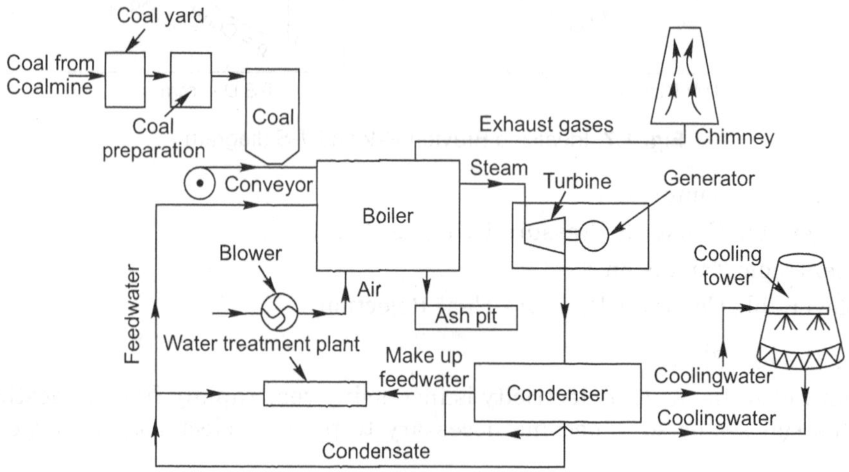

Fig. 1.8. Steam/thermal power plant layout

The coal is received in coalstorage yard of power station. The coal is sized by crushers. It is then stored in stockyard from which it is then transferred to the boiler furnace by means of conveyors.

The coal is burnt in the boiler furnace and hence, ash is formed in the boiler furnace. This hot, dusty and poisonous gases accompanying ash is transferred to the ash storage where it usually quenched to reduce the temperature, corrosion and dust content.There are different methods employed for the disposal of ash. They are hydraulic system, water jetting, ash sluice ways, pneumatic system, etc.

In hydraulic system, which is used in large power plants, ash falls from the furnace grate into high velocity water stream. It is then carried to the slumps.

2. Feedwater and steam flow circuit: It includes feed pump, economizer, boiler drum, super heater, turbine condenser etc. Feedwater is pumped to the economizer from the hot well. It is preheated by the flue gases in the economizer and then supplied to the boiler drum where heat is transferred to the water from the combustion of the coal. Hence, this water gets converted into the steam. This steam produced in the boiler is sent to a super heater where it is super heated by the flue gases. This super heated steam is then expanded in a steam turbine to do mechanical work to produce electric power.

The expanded steam is then passed through the condenser where it is condensed into water and recirculated to hot well.

3. Air and flue gas circuit: It includes forced draught fan, air preheated, boiler furnace, super heater, economizer, dust collector, induced draught fan, chimney, etc. The forced draught fan supplies the air to the air preheated from the atmosphere by the action of a forced draught fan.

This air is preheated by the flue gases in the air preheated. This preheated air is supplied to the furnace to aid the combustion of the fuel due to the combustion of the fuel. Flue (hot) gases are formed.

The flue (hot) gases pass over boiler tubes where wet steam is generated and pass over super heater where the wet steam is super heated. The flue gases then pass through economizer to heat the feedwater. It is then passed through the air preheated to preheat incoming air. Finally it is sent to the atmosphere through chimney after the dusts in the flue gases are arrested in the dust collector.

4. Cooling water circuit: It includes a pump, condenser, cooling tower, etc. The steam from the turbine is condensed in the condenser. The steam is condensed by loosing its latent heat to the circulating cold water. This heated circulating water is then taken to the cooling tower.

In the cooling tower, the water is sprayed in the form of droplets through nozzles. The atmospheric air enters the cooling tower and removes the heat from the water. This cooled water is collected in the bottom cooling pond and then again circulated to the condenser and then cooling tower through the pump.

Merits of Steam/Thermal Power Plant

- Coal used as fuel is cheap.
- Can be installed anywhere near fuel source and water source. Requires less construction space.
- Setting-up cost is lower compared to hydroelectric power plant. Construction time is less when compared to hydroelectric power plant.

Demerits of Steam/Thermal Power Plant

- Pollutes air in the form of smoke.
- Operating cost (cost of power generation) is more compared to the other types of power plants.
- High cost of transmission and distribution if the plant is located far away from load centres.
- Availability of fuel (coal) in the long-run is less.
- Cannot be used during peakload.
- Stand-by losses are heavy since boiler continues to work even after the turbine is switched off.

1.6 MAJOR COMPONENTS OF THERMAL (STEAM) POWER PLANT

Thermal power plant is nothing but the plant at which power (electricity) is generated by using steam as a working fluid. Steam has the advantage that it can be produced from water which is available in abundance. It does not react much with the material of the equipment of power plant. The high kinetic energy of steam is used to drive the steam turbines (prime mover) steam engines etc. Thermal power plant is most suited for easy power generation. About 60% of the total power generated in our county comes from thermal power plant. The study of major components of thermal power plant is vital role in thermal engineering. The major components are shown in following schematic diagram of a thermal (steam) power plant for more understanding:

1. Boiler
2. Turbine
3. Condenser
4. Pump

In this chapter, we shall discuss about the major components of a thermal power plant in the following sections.

1.6.1 Boiler

1. Introduction

A boiler also known as steam generator is a closed vessel generally made of high quality steel in which steam is generated by heating water. Required heat is produced by the combustion of solid, liquid or gaseous fuel. The vapourized fluid *i.e.*, steam, leaves the boiler for use in various processes or heating applications. The steam produced is used for

1. Producing mechanical work
2. Heating residential and industrial building
3. Performing certain processes in the sugarmills, chemical and textile industries.

Uses of Boilers

Boilers are used:

1. for producing steam at high pressure in a thermal power plant.
2. in the processing of sugar, chemical, textile and food items, etc. at low pressure.
3. for heating the residential and industrial buildings at very low pressure.
4. To provide steam for locomotion in trains, ships and boats.

Requirements of Good Boiler

The requirements of a good boiler are the following:

1. A good boiler should produce maximum quantity of steam with minimum fuel usage.
2. It should be easy to start.
3. It should be easy to handle.
4. It should require minimum attention during operation.
5. It should be well suited for all fluctuations in load.
6. It should involve low initial and operation cost.
7. It should satisfy all safety regulations as prescribed in the Boilers Act.

Further the boiler should fulfil the following requirements.

(*i*) **Safety.** The boiler should be safe under operating conditions.

(*ii*) **Accessibility.** The various parts of the boiler should be accessible for repair and maintenance.

(*iii*) **Capacity.** The boiler should be capable of supplying steam according to the requirements.

(*iv*) **Efficiency.** To permit efficient operation, the boiler should be able to absorb a maximum amount of heat produced due to burning of fuel in the furnace.

(*v*) It should be simple in construction and its maintenance cost should be low.

(*vi*) Its initial cost should be low.

(*vii*) The boiler should have no joints exposed to flames.

(*viii*) The boiler should be capable of quick starting and loading.

The performance of a boiler may be measured in terms of its evaporative capacity also called power of a boiler. It is defined as the amount of water evaporated or steam produced in kg per hour. It may also be expressed in kg per kg of fuel burnt or kg/hr/m^2 of heating surface.

2. Classification of Boilers

The boilers can be classified as:

- Water tube boiler and
- Fire tube boiler

3. Water Tube Boiler

Water circulation: In water tube boiler, water circulates inside the tubes which are surrounded by hot gases from the furnace.

Steam pressure: Water tube boiler generates steam at a pressure up to 165 bars.

Floor area requirement, flexibility and generation rate: In water tube boiler, the floor area requirement for the generation of steam is less for a given power. The boiler can respond quickly to change in steam demand. It generates steam as high as 450 tones per hour.

Operating cost: Operating cost of water tube boiler is high.

Efficiency: The high rate of movement of water in the tubes enhances the rate of heat transfer resulting in high efficiency. It is used in large power plants.

Examples of water tube boiler: Babcock and Wilcox boiler and Benson boiler.

Various advantages of water tube boilers are as follows:

(*i*) High pressure of the order of 140 kg/cm^2 can be obtained.

(*ii*) Heating surface is large. Therefore steam can be generated easily.

(*iii*) Large heating surface can be obtained by use of large number of tubes.

(*iv*) Because of high movement of water in the tubes the rate of heat transfer becomes large resulting into a greater efficiency.

4. Fire Tube Boiler

Water circulation: In fire tube boiler, the hot gases from the furnace pass through the tubes which are surrounded by water.

Steam pressure: A fire tube boiler generates steam at a pressure up to 24 bars.

Floor area requirement, flexibility and generation rate: In fire tube boiler, the floor area requirement for the generation of steam is more for a given power. The boiler cannot respond quickly to change in steam demand. It generates only a limited quantity of steam.

Operating cost: Operating cost of fire tube boiler is low.

Efficiency: The movement of water around the tubes is less. This leads to very low rate of heat transfer resulting in low efficiency.

Examples of fire tube boiler: Lancashire boiler, Cochran boiler and locomotive boiler.

Various advantages of fire tube boilers are as follows:

(*i*) Low cost

(*ii*) Fluctuations of steam demand can be met easily

(*iii*) It is compact in size.

5. Boiler Mountings and Accessories

Boiler mountings: Fittings which are necessary for the safety and control of boiler are known as boiler mountings.

- Fittings that are essential from the safety point of view are the following:
  1. Water level indicators
  2. Safety valves
  3. Combined high steam and low water safety valve
  4. Fusible plug
- Fittings which are essential from the control point of view are the following:
  1. Pressure gauge
  2. Junction valve or stop valve
  3. Feed check valve
  4. Blow-off cock
  5. Man hole and mud box

Boiler accessories: Fittings which are provided to increase the efficiency of the boiler and help in the smooth working of the boiler are known as boiler accessories. The important accessories are the following:

1. Super heater
2. Economizer
3. Air pre-heater
4. Feed pump

Water level indicator: Water level indicator shows the boiler operator the level of fluid in the boiler. It is also known as sight glass, water gauge or water column.

Safety valve: Safety valve is used to relieve pressure and prevent possible explosion of the boiler.

Fusible plug: Fusible plug acts as a last-resort safety device in the event of the water level falling dangerously low. Fusible plug is a threaded metal plug, usually of bronze, brass or gunmetal, with a tapered hole drilled completely through its length. This hole is sealed with a metal of low melting point, usually lead or tin. It is screwed into the crown sheet of a boiler, typically extending about 25 mm into the water space above. When the top of the plug is out of the water, it gets overheated and the lead core melts away. The resulting noisy release of steam into the firebox warns the operators of the danger before the top of the firebox itself runs completely dry.

Super heater: A super heater is a device in a boiler that heats the steam generated by the boiler again thereby increasing its thermal energy. Super heater increases the efficiency of the boiler. Steam which has been superheated is logically known as superheated steam.

Economizer: Economizer or heat exchanger is a mechanical device intended to reduce energy consumption or to perform another useful function like preheating a fluid.

Air pre-heater: An air pre-heater or air heater is a general term to describe any device designed to heat air before another process with the primary objective of increasing the thermal efficiency of the process. The purpose of the air pre-heater is to recover the heat from the boiler flue gas thus increasing the thermal efficiency of the boiler by reducing the useful heat lost in the flue gas.

6. Babcock and Wilcox Boiler

This is a low pressure water tube boiler. It is used in low capacity power plants.

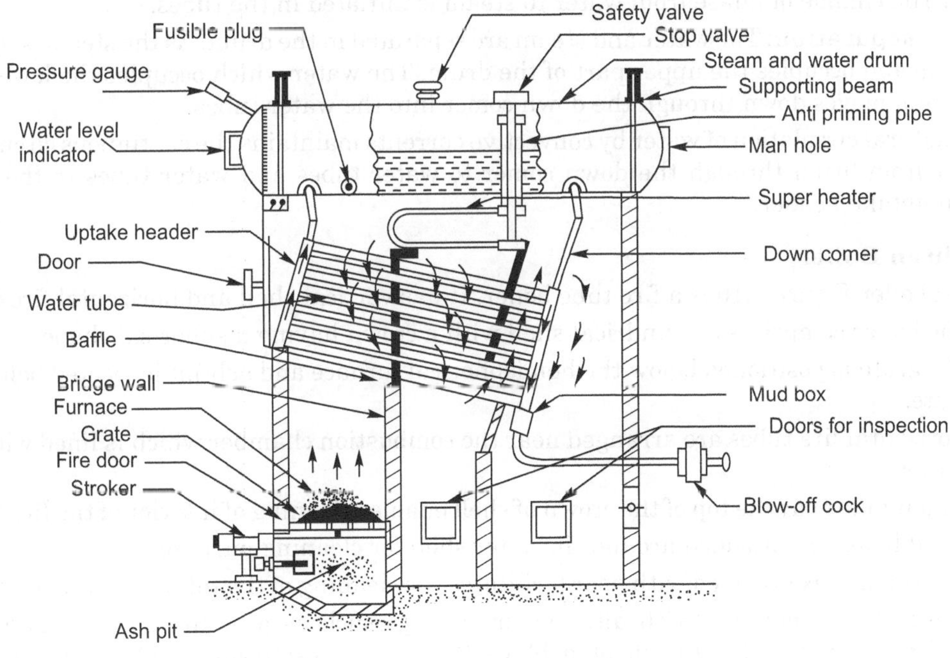

Fig. 1.9. Babcock and Wilcox boiler

Construction

Babcock and Wilcox boiler is a straight tube, stationary type water tube boiler. In this boiler, water circulates inside the tubes which arc surrounded by hot gases from the furnace. It consists of a steam and water drum which is connected by short tubes with uptake header and down comer.

The water tubes which connect the uptake header to the down comer are inclined at an angle of 15° to the horizontal as shown in Figure 1.9. The tubes are arranged in equal spacing in order to facilitate the hot gases to pass over the tubes without any obstruction in the passage. This arrangement increases the heat transfer rate. A mud box fitted to the down comer is used to collect the impurities and mud particles from the water and they are blown-off periodically by means of blow-off cock.

Fittings or devices are used to increase the efficiency of the boiler and help in the smooth working of the plant. Devices which are necessary for the safety and control called boiler mountings are also used in the boiler.

Flue gas line: The furnace is arranged below the uptake header. Coal is fed to the grate through the fire door and fired. Two fire brick baffles are arranged in such a way that the hot gases from the grate are made to pass upward and downward alternately. This facilitates the water tubes to absorb maximum amount of heat energy from the flue gas and finally, the flue gas exits out through the smoke chamber or chimney.

Water line: Water is passed through the tubes. The water tubes are positioned at an angle and the inclination of tubes enhances the water circulation. The water tubes which are nearer to the furnace get heated faster compared to the remaining tubes. Due to absorption of heat, the water in the tubes gets heated to a very high temperature which leads to the decrease in density of water. The water with decreased density rises up into the drum through the uptake header. The change of phase from water to steam is initiated in the tubes.

Steam separation: The water and steam are separated in the drum. As the steam is lighter, it rises up and occupies the upper part of the drum. The water which occupies the lower part of the drum moves down through the down comer into the water tubes.

The natural circulation of water by convective currents maintains the continuous circulation of water from drum through the down comer to water tubes and water tubes to the drum through uptake header.

7. Cochran Boiler

Cochran boiler Figure 1.10 is a fire tube boiler. It has vertical shell and horizontal fire tubes:

- The boiler comprises a cylindrical shell with a crown having a spherical shape.
- The grate is positioned below the hemispherical furnace and ash-pit is located below the grate.
- Horizontal fire tubes are arranged near the combustion chamber which is lined with fire bricks.
- A man-hole near the top of the crown of shell enables cleaning of interior of the fire tubes.
- Hand holes are provided around the outer shell for cleaning purpose.
- The smoke box provided with door facilitates the cleaning of the interior of the fire tubes.
- The boiler is mounted with different mountings such as pressure gauge, water level indicator, safety valve, fusible plug, blow-off cock, steam stop valve and feed check valve.

Working Principle

The coal is fed into the grate through the fire door and ignited. The ash formed is collected in the ash-pit located just below the grate and it is removed manually. The hot gases produced due to combustion of coal in the combustion chamber flow through the horizontal fire tubes. The hot gases passing through the fire tubes transfer a large portion of the heat to the water surrounding the tubes by convection. The flue gases coming out of the fire tubes are discharged to the atmosphere through the chimney.

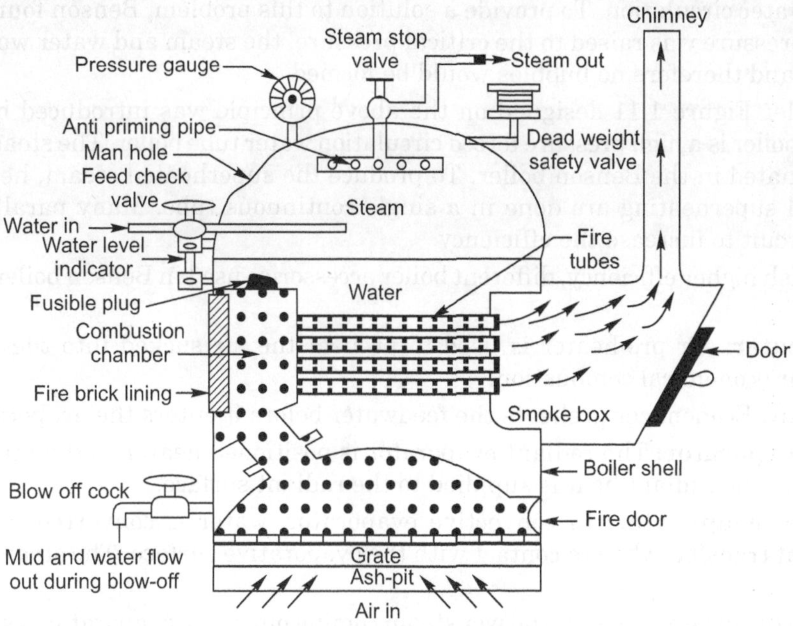

Fig. 1.10. Cochran boiler

High Pressure Boilers

High pressure boilers are used in power plants to generate steam at a high pressure. The modem high pressure boilers used for power generation have steam capacities ranging from 40 to 650 tones/h with pressure up to 200 bar and maximum steam temperature of about 600°C.

The special features of high pressure boilers are the following:

1. Provision of tubings and drums
2. Forced water circulation
3. Evaporation of water above critical pressure resulting in the saving of latent heat
4. Use of super saturated steam for heating water.

Advantages of High Pressure Boiler

1. Overheating and development of thermal stresses are avoided because of uniform heating of all parts.
2. The high velocity of circulation of fluid reduces the formation of scales to a great extent.
3. The rapid changes of load can be easily controlled.
4. The high pressure boiler is light and compact.
5. The efficiency of high pressure boiler is fairly high.

Examples of High Pressure Boiler

(*i*) Benson boiler (*ii*) Lamont boiler
(*iii*) Loeffler boiler

8. Benson Boiler

In general, the rate of heat transfer from the flue gas to water is seriously affected by the presence of steam bubbles that are in contact with the tube. The release of these bubbles disturbs the water circulation. To provide a solution to this problem, Benson found that if the boiler steam pressure was raised to the critical pressure, the steam and water would have the same density and therefore no bubbles would be formed.

Benson boiler Figure 1.11 designed on the above principle was introduced by Benson in 1922. Benson boiler is a high pressure forced circulation water tube boiler. The steam separating drum is eliminated in the Benson boiler. To produce the superheated steam, heating, steam formation and superheating are done in a single continuous tube. Many parallel tubes are used in the circuit to increase the efficiency.

To accomplish higher efficiency, different boiler accessories used in Benson boiler are detailed below:

Air pre-heater: Air pre-heater is used to preheat the air sucked into the system from atmosphere for economical combustion.

Economizer: Economizer preheats the feedwater before it enters the evaporator.

Radiant evaporator: The radiant evaporator is positioned nearer to the furnace. Due to the combustion, the radiant heat is supplied to the radiant surface.

Convective evaporator: In convective evaporator, water is converted into steam by convective heat transfer, when in contact with the evaporative surface. The steam produced is wet.

Super heater: In super heater, the wet steam coming out of the evaporator is super-heated.

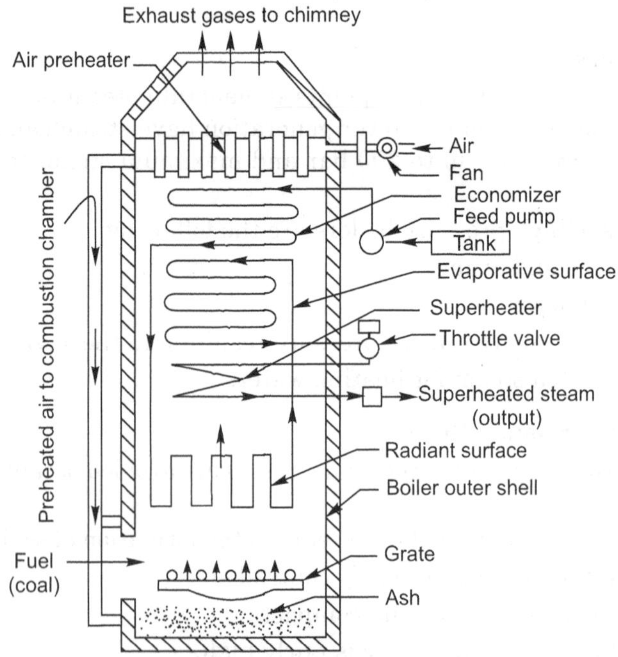

Fig. 1.11. Benson boiler

The superheated steam is the output of the boiler system which can be utilized for works such as running the steam turbine in the case of power generation.

A large amount of power is consumed by the feed pump to raise the pressure above the critical pressure which leads to lower efficiency of the plant. So, by using pressure lower than the critical pressure, improved efficiency is possible.

At normal load, Benson boiler has an operating pressure of 210 bar with 13.5 tones per hour evaporating capacity at 405°C superheating temperature.

Advantages of Benson Boiler

1. There is no steam separating drum, thus simpler in design.
2. There is no circulating pump and down comers.
3. Easy to install.
4. Evaporation rate is high.
5. Light in weight.
6. Higher specific output with high safety factor.

Disadvantages of Benson Boiler

1. Corrosion takes place in the tubes. Hence, special care is required to monitor this corrosion.
2. Chances of overheating of surfaces in case of insufficient water supply.
3. Effective monitoring and maintenance are required as the storage capacity is less.
4. Special arrangements are required to remove the salt and solids formed during the evaporation of water in the tubes.

9. Lamont Boiler

Lamont boiler Figure 1.12 was introduced by Lamont in 1925. Lamont boiler is a high pressure, forced circulation water tube boiler.

Functions of Different Components of Lamont Boiler

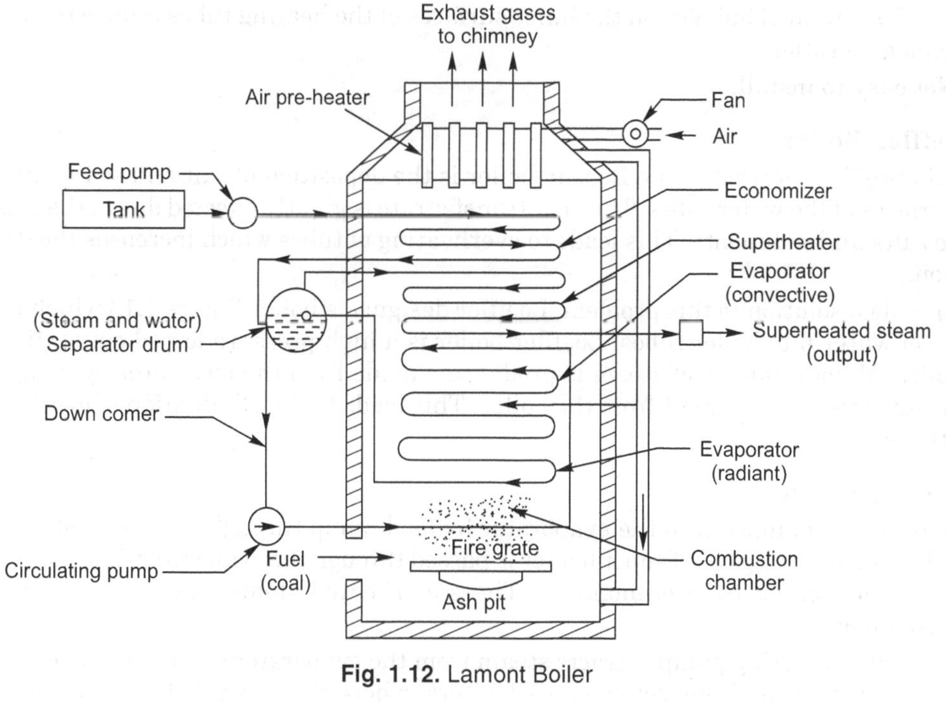

Fig. 1.12. Lamont Boiler

Feedwater line: The feedwater from the tank/well is supplied to the economizer. From the economizer, it passes through separator drum (which separates steam and water) that contains a feed regulator which controls the speed of feed pump.

Combustion: Combustion is made economical by forced air preheating by the flue gases.

Economizer: Economizer receives heat from hot combustion gases and supplies sensible heat to the feedwater.

Circulating pump: Feedwater after passing through economizer and boiler drum flows to a circulating pump. Circulating pump rises the water pressure (to about 3.5 bar) above the drum pressure, facilitating flow through flow controlling orifices and small diameter tubing. The pump consumes 0.5–0.6% of boiler output.

Super heater: The steam is lead to the super heater tubes in which the steam gets superheated.

Radiant evaporator: The radiant evaporator is positioned nearer to the furnace. The radiant heat is supplied to the radiant surface.

Convective evaporator: In convective evaporator, the water is converted into steam by convective heat transfer, when in contact with the evaporative surface. The steam produced is wet. The superheated steam then passes through the steam stop valve to the turbine.

Maintaining of water level: The water level in the drum is maintained at constant level by supplying feedwater equivalent to the quantity of steam blown-off.

Output capacity: At normal load, Lamont boiler has an operating pressure of 120 bar with 40 to 50 tones per hour of superheated steam generation capacity at 520 °C superheating temperature.

Disadvantages of Lamont Boiler

1. The formation of bubbles on the inner surfaces of the heating tubes reduce the heat flow and steam generation.
2. Not easy to install.

10. Loeffler Boiler

The main problem noticed in the Benson boiler is the deposition of salt and sediments on the inner surfaces of the water tubes. The heat transfer rate is greatly affected due to the deposition of these salts and sediments. This leads to overheating of tubes which increases the danger of explosion.

To provide a solution to this problem, Loeffler designed a boiler Figure 1.13 which prevents the flow of water into boiler tubes. Loeffler boiler is a high pressure forced circulation water tube boiler. Major quantity of steam is produced outside from the feedwater by using part of superheated steam coming out from the boiler. This leads to the elimination of overheating of water tubes.

Working Principle

The feedwater is pumped into the evaporator by feed pump through the economizer. Nearly 65% of the steam coming out of super heater is passed through the evaporator drum to evaporate the feedwater coming from economizer. The superheated steam converts the water into saturated steam.

The steam circulating pump extracts steam from the evaporator drum and forces it to pass through the radiant and convective super heaters, where the steam is heated to the desired temperature. About 35% of the steam from the super heater is passed to the prime mover and the remaining steam (65%) flows to the drum.

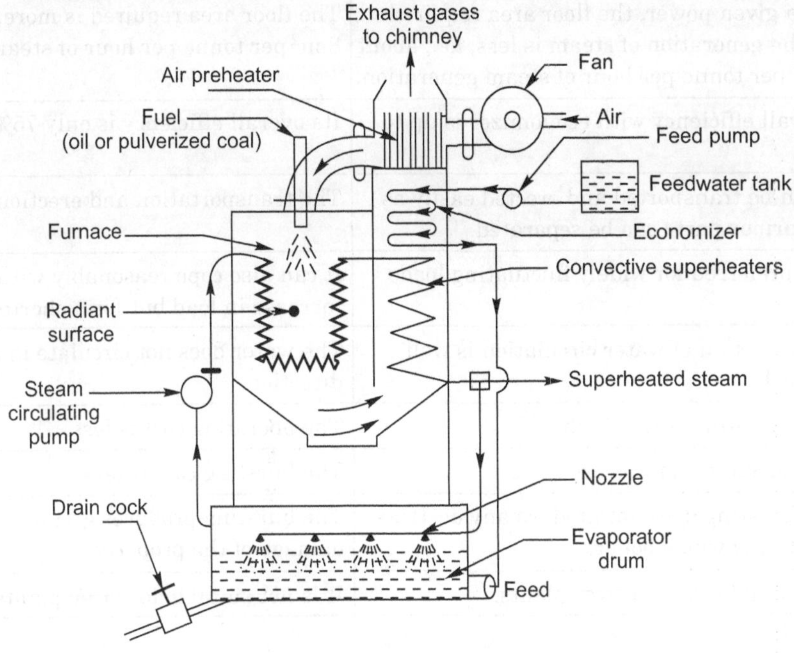

Fig. 1.13. Loeffler boiler

At normal load, Loeffler boiler has an operating pressure of 140 bar with 94.5 tonnes per hour evaporating capacity at about 482°C superheating temperature.

Advantages of Loeffler Boiler

1. Simple in design and easy to install.
2. Steam circulation is well controlled.
3. Evaporation rate is high and higher specific output with high safety factor.
4. Formation of boiler scales is absent.
5. It responds quickly to variations in load.

Disadvantage of Loeffler Boiler

Special arrangements are required to monitor and maintain the nozzles which distribute the superheated steam to the evaporator drum. Otherwise, there will be noise and need for elaborate-miming.

11. Comparison between Water Tube and Fire Tube Boiler

S. No.	*Water tube boiler*	*Fire tube boiler*
1.	The water circulates inside the tubes which are surrounded by hot gases from the furnace.	The hot gases from the furnace pass through the tubes which are surrounded by water.
2.	It generates steam at a higher pressure up to 165 bars.	It can generate steam only up to 24.5 bars.
3.	The rate of generation of steam is high, *i.e.*, up to 450 tonnes per hour.	The rate of generation of steam is low, *i.e.*, up to 9 tonnes per hour.

4.	For a given power, the floor area required for the generation of steam is less, *i.e.*, about 5 m^2 per tonne per hour of steam generation.	The floor area required is more, *i.e.*, about 8 m^2 per tonne per hour of steam generation.
5.	Overall efficiency with economizer is up to 90%.	Its overall efficiency is only 75%
6.	It can be transported and erected easily as its various parts can be separated.	The transportation and erection is difficult.
7.	It is preferred for widely fluctuating loads.	It can also cope reasonably with sudden increase in load but for a shorter period.
8.	The direction of water circulation is well defined.	The water does not circulate in a definite direction.
9.	The operating cost is high.	The operating cost is less.
10.	The bursting chances are more.	The bursting cost is less.
11.	The bursting does not produce any destruction to the whole boiler.	The bursting produces greater risk to the damage of the property.
12.	It is used for large power plants.	It is not suitable for large plants.

1.6.2 Turbines

1. Introduction - Steam Turbine

A steam turbine is a prime mover in which rotary motion is obtained by the gradual change of momentum of the steam. We have already discussed that in a reciprocating steam engines, the steam acts on the piston, as a load or weight, *i.e.*, the action of steam is static. The statical pressure of steam gives to and fro motion to the piston, and conversion of energy takes place through crank and connecting rod.

In a steam turbine, the force exerted on the blades is due to the velocity of steam. This is due to the fact that the curve blades by changing the direction of steam receive a force or impulse. The action of steam in this case is said to be dynamic. Thus, the dynamical pressure of steam rotates the vanes, buckets or blades directly. The turbine blades are curved in such a way that the steam directed upon them enters without shock, though there is always some lost of energy by the friction upon the surface of blades. In general, a steam turbine, essentially, consists of the following two parts:

1. The nozzle in which the heat energy of high pressure steam is converted into kinetic energy, so that the steam issues from the nozzle with a very high velocity.

2. The blades which change the direction of steam issuing from the nozzle, so that a force acts on the blades due to change of momentum and propel them.

Thus, the basic principle of operation of a steam turbine is the generation of high velocity steam jet by the expansion of high pressure steam and then conversion of kinetic energy, so obtained into mechanical work on rotor blades.

Classification of Steam Turbines

The steam turbines may be classified into the following types:

1. According to the mode of steam action

   (*i*) Impulse turbine, and (*ii*) Reaction turbine.

2. According to the direction of steam flow
   (*i*) Axial flow turbine, and (*ii*) Radial flow turbine.
3. According to the exhaust condition of steam
   (*i*) Condensing turbine, and (*ii*) Non-condensing turbine.
4. According to the pressure of steam
   (*i*) High pressure turbine, (*ii*) Medium pressure turbine, and
   (*iii*) Low pressure turbine.
5. According to the number of stages
   (*i*) Single stage turbine, and (*ii*) Multi-stage turbine.

2. Impulse Turbine

Introduction

An impulse turbine, as the name indicates, is a turbine which runs by the impulse of steam jet. In this turbine, the steam is first made to flow through a nozzle. Then the steam jet impinges on the turbine blades (which are curved buckets) and are mounded on the circumference of the wheel is shown in Figure 1.14. The steam jet after impinging glides over the concave surface of the blades and finally leave the turbine.

Note: The action of the jet of steam, impinging on the blades, is said to be an impulse and the rotation of the rotor is due to the impulsive force of the steam jets.

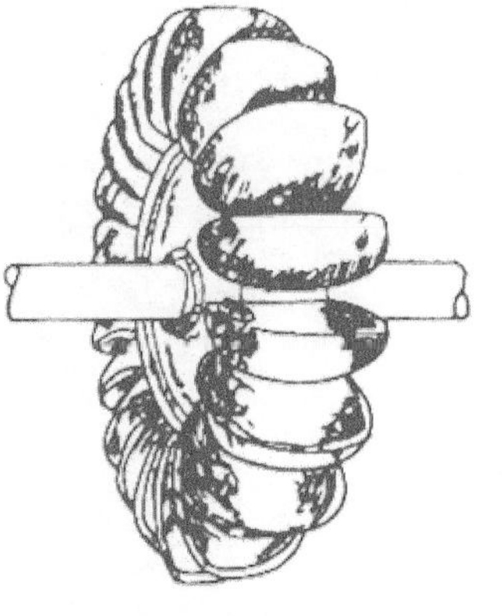

Fig. 1.14. Impulse turbine motor

De-Level Impulse Turbine

A De-Level turbine is the simplest type of impulse steam turbine, and is commonly used. It has the following main components:

1. Nozzle. It is a circular guide mechanism, which guides the steam to flow at the designed direction and velocity. It also regulates the flow of steam. The nozzle is kept very close to the blade, in order to minimize the losses due to windage.

2. Runner and blades. The runner of a De-Level impulse turbine essentially consists of a circular disc fixed to a horizontal shaft. On the periphery of the runner, a number of blades are fixed uniformly. The steam jet impinges on the buckets, which move in the direction of the jet. This movement of the blades makes the runner to rotate. The surface of the blades is made very smooth to minimize the frictional losses. The blades are generally made of specials steel alloys. In most of the cases, the blades are bolted to the runner disc. But sometimes the blades and disc are cast as a single unit.

It has been experienced that all the blades do not wear out equally with the time. A few of them get worn out and damaged early and need replacement. This can be done only if the blades are bolted to the disc.

3. Casing. It is an air-tight metallic case, which contains the turbine runner and blades. It controls the movement of steam from the blades to the condenser, and does not permit it to move into the space. Moreover, it is essential to safeguard the runner against any accident.

Pressure and Velocity of Steam in an Impulse Turbine

The pressure of steam jet is reduced in the nozzle and remains constant while passing through the moving blade. The velocity of steam is increased in the nozzle, and is reduced while passing through the moving blades. Figure 1.15 shows the pressure and velocity graphs of the steam in a simple impulse turbine while it flows in the nozzle and blades. The pressure graph 1-2-3-4 represents steam pressure at entrance of the nozzle, exist of the nozzle, entrance of the blades and exit of the blades respectively. Similarly, velocity graph 5-6-7-8 represents the velocity of steam at entrance of the nozzle, exit of the nozzle, entrance of the blade and exit of the blades respectively.

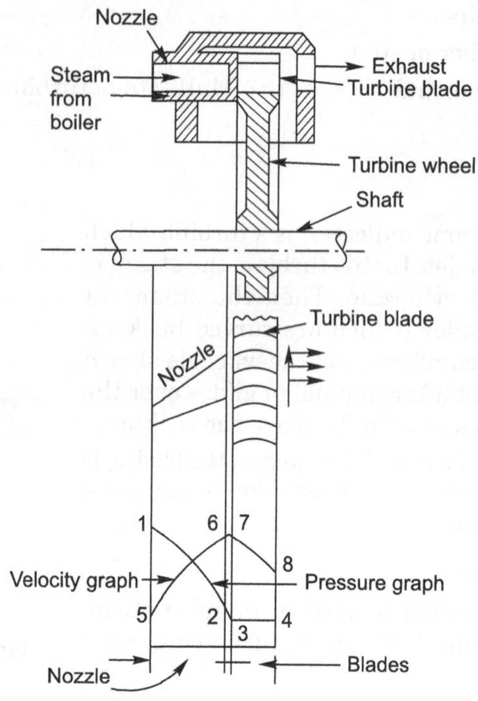

Fig. 1.15. Pressure and Velocity of Steam in an Impulse Turbine

3. Reaction Turbine

Introduction

In a reaction turbine, the steam enters the wheel under pressure and flows over the blades. The steam, while gliding, propels the blades and make them to move. As a matter of fact, the turbine runner is rotated is rotated by the reactive forces of steam jets. The backward motion of the blades is similar to the recoil of a gun. It may be noted that an absolute reaction turbine is rarely used in actual practice.

Fig. 1.16. Reaction Turbine-runner

Parson's Reaction Turbine

A Parson's turbine is the simplest type of reaction steam turbine, and is commonly used. It has the following main components:

1. Casing. It is an air-tight metallic case, in which the steam from the boiler, under a high pressure and temperature, is distributed around the fixed blades (guide mechanism) in the casing. The casing is designed in such a way that the steam enters the fixed blades with a uniform velocity.

2. Guide mechanism. It is a mechanism, made up with the help of guide blades, in the form of a wheel. This wheel is, generally, fixed to the casing; that is why these guide blades are also called fixed blades. The guide blades are properly designed in order to:

(*a*) Allow the steam to enter the runner without shock. This is done by keeping the relative velocity at inlet of the runner tangential to the blade angle.

(*b*) Allow the required quantity of steam to enter the turbine. This is done by adjusting the openings of the blade.

The guide blades may be opened or closed by rotating the regulating shaft, thus allowing the steam to flow according to the need. The regulating shaft is operated by means of a governor whose function is to govern the turbine (*i.e.*, to keep the speed constant at varying loads).

3. Turbine runner. The turbine runner of a Parson's reaction turbine essentially consists of runner blades fixed to a shaft or rings, depending upon the type of turbine. The blades, fixed to the and leave the runner without shock, as shown in Figure 1.16. The surface of the turbine runner is made very smooth to minimize the frictional losses. The turbine runner is, generally, cast in one piece. But sometimes, it is made up of separate steel plates welded together.

4. Draft tube. The steam, after passing through the runner, flows into the condenser through a tube called draft tube. It may be noted that if this tube is not provided in the turbine, then the steam will move freely and will cause steam eddies.

Pressure and Velocity of Steam in Reaction Turbine

It will be interesting to know that the pressure in a reaction turbine is reduced in the fixed blades as well as in moving blades. The velocity of steam is increased in the fixed blades, and is reduced while passing through the moving blades.

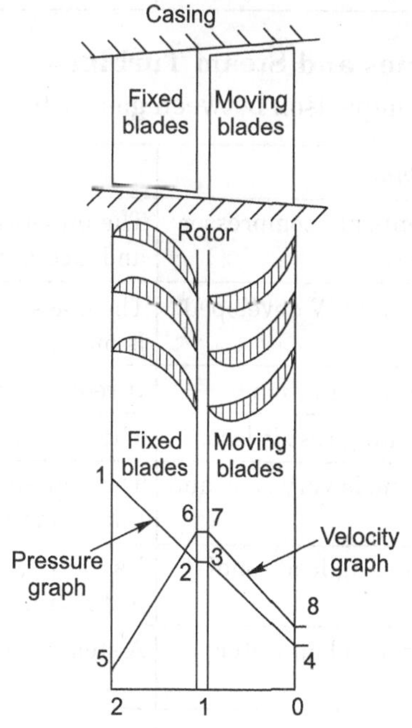

Fig. 1.17. Pressure and Velocity of Steam in Reaction turbine

Figure 1.17 shows the pressure and velocity graphs of the steam while it flows in the fixed and moving blades of a reaction turbine. The pressure graph 1-2-3-4 represents steam pressure at entrance of the fixed blades, exit of the moving blades, entrance of the moving blades and exit of the moving blades respectively. Similarly, velocity graph 5-6-7-8 represents the velocity of steam at entrance of the fixed blades, exit of the fixed blades, entrance of the moving blades and exit of the moving blades respectively.

4. Comparison between Impulse Turbine and Reaction Turbine

Following are the few points of comparison between an impulse turbine and a reaction turbine:

S. No.	*Impulse turbine*	*Reaction turbine*
1.	The steam flows through the nozzles and impinges on the moving blades.	The steam flows first through guide mechanism and then through the moving blades.
2.	The steam impinges on the buckets with kinetic energy.	The steam glides over the moving vane with pressure and kinetic energy.
3.	The steam may or may not be admitted over the whole circumference.	The steam must be amitted over the whole circumference.
4.	The steam pressure remains constant during its flow through the moving blades.	The steam pressure is reduced during its flow through the moving blades.
5.	The relative velocity of steam while gliding over the blades remains constant (assuming no friction).	The relative velocity of steam while gliding over the moving blades increases (assuming no friction).
6.	The blades are symmetrical. The number of stages required is less for the same power developed.	The blades are not symmetrical. The number of stages required is more for the same power developed.

5. Comparison of Gas Turbines and Steam Turbines

Following are the points of comparison between gas turbines and steam turbines:

S. No.	*Gas turbines*	*Steam turbines*
1.	The important components are compressor and combustion chamber.	The important components are steam boiler and accessories.
2.	The mass of gas turbine per kW developed is less.	The mass of steam turbine per kW developed is more.
3.	It requires less space for installation.	It requires more space for installation.
4.	The installation and running cost is less.	The installation and running cost is more.
5.	The starting of gas turbine is very easy and quick.	The starting of steam turbine is difficult takes long time.
6.	Its control, with the changing load conditions, is easy.	Its control, with the changing load conditions, is difficult.
7.	A gas turbine does not depend on water supply.	A steam turbine depends on water supply.
8.	Its efficiency is less.	Its efficiency is higher.

6. Comparison of Gas Turbines and I.C. Engines

Following are the points of comparison between gas turbines and I.C. engines:

S. No.	*Gas turbines*	*I.C. engines*
1.	The mass of gas turbine per kW developed is less.	The mass of an I.C. Engine per kW developed is more.
2.	The installation and running cost is less.	The installation and running cost is more.
3.	Its efficiency is less.	Its efficiency is higher.
4.	The balancing of gas turbine is perfect.	The balancing of I.C. Engine is not perfect.
5.	The torque produced is uniform. Thus no flywheel is required.	The torque produced is uniform. Thus fly-wheel is necessary.
6.	The lubrication and ignition systems are simple.	The lubrication and ignition systems are difficult.
7.	It can be driven at a very high speed.	It cannot be driven at a very high speed.
8.	The pressure used are very low (about 5 bar).	The pressure used are high (above 60 bar).
9.	The exhaust of a gas turbine is free from smoke and less polluting.	The exhaust of an I.C. Engine is more polluting.
10.	They are very suitable for air crafts.	They are less suitable for air crafts.
11.	The starting of a gas turbine is not simple.	The starting of an I.C. Engine is sample.

1.6.3 Condensers

1. Introduction

A steam condenser is a closed vessel into which the steam is exhausted, and condensed after doing work in an engine cylinder or turbine. A steam condenser has the following two objects: The steam condensing plant is shown in Figure 1.18.

1. The primary object is to maintain a low pressure (below atmospheric pressure) so as to obtain the maximum possible energy from steam and thus to secure a high efficiency.

2. The secondary object is to supply pure feedwater to the hot well, from where it is pumped back to the boiler.

Note: The low pressure is accompanied by low temperature and thus all condensers maintain a vacuum under normal conditions. The condensed steam is called condensate. The temperature of condensate is higher on leaving the condenser than that of circulating water at inlet. It is thus obvious, that the condensate will have a considerable liquid heat.

2. Classification

The steam condensers may be broadly classified in to the following two types, depending upon the way in which the steam is condensed:

A. Jet condensers or mixing type condensers, and

B. Surface condensers or non-mixing type condensers.

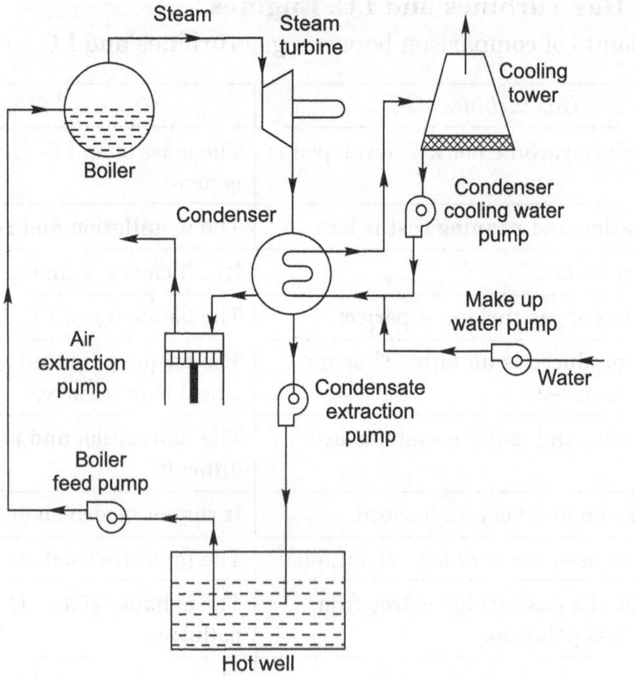

Fig 1.18. Steam Condensing plant

A. Jet Condensers

These days, the jet condensers are seldom used because there is some loss of condensate during the process of condensation and high power requirements for the pumps used. Moreover, the condensate cannot be used as feedwater to the boiler as it is not free from salt. However, jet condensers may be used at places where water of good quality is easily available in sufficient quantity.

Types of Jet Condensers

The jet condensers may be further classified, on the basis of the direction of flow of the condensate and the arrangement of the tubing system, in to the following four types:

1. Parallel flow jet condenser,
2. Counter flow or low level jet condenser,
3. Barometric or high level jet condenser and
4. Ejector condenser,

These condensers are discussed, in detail, in the following pages.

Parallel Flow Jet Condensers

In parallel flow jet condensers, both the steam and water enter at the top, and the mixture is removed from the bottom.

The principle of this condenser is shown in Figure 1.19. The exhaust steam is condensed when it mixes up with water. The condensate, cooling water and air flow downwards and are removed by two separate pumps known as air pump and condensate pump. Sometimes, a single pump known as wet air pump, is also used to remove both air and condensate. But the former gives a greater vacuum. The condensate pump delivers the condensate to the hot well, from where surplus water flows to the cooling water tank through an overflow pipe.

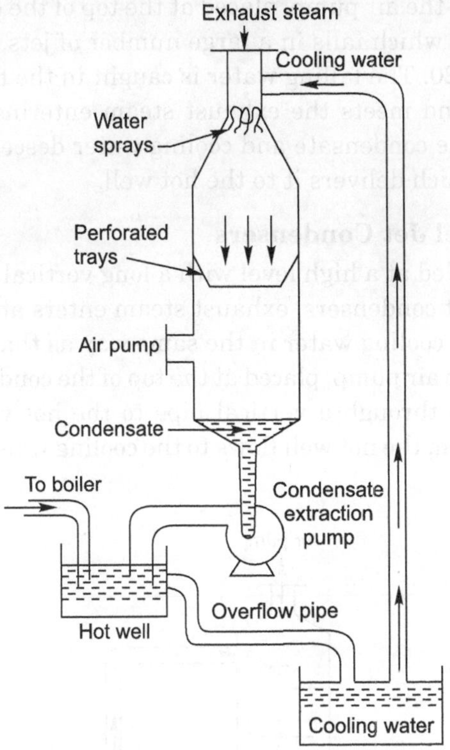

Fig. 1.19. Parallel Flow Jet Condenser

Counter Flow or Low Level Jet Condensers

In counter flow or low level jet condensers, the exhaust steam enters at the bottom, flows upwards and meets the down coming cooling water.

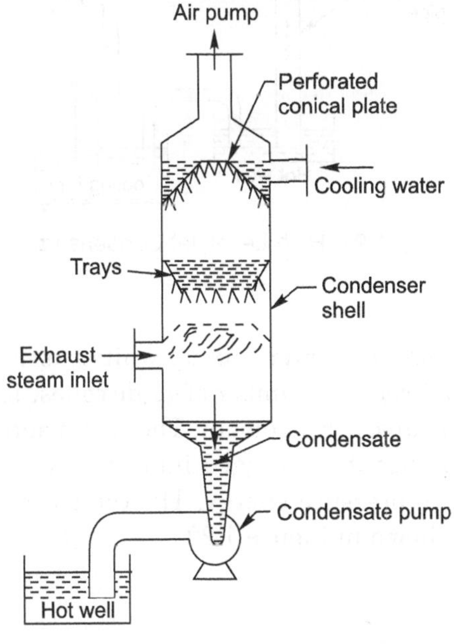

Fig. 1.20. Counter Flow Jet Condenser

The vacuum is created by the air pump, placed at the top of the condenser shell. This draws the supply of cooling water, which falls in a large number of jets, through perforated conical plate as shown in Figure 1.20. The falling water is caught in the trays, from which it escapes in a second series of jets and meets the exhaust steam entering at the bottom. The rapid condensation occurs, and the condensate and cooling water descends through a vertical pipe to the condensate pump, which delivers it to the hot well.

Barometric or High Level Jet Condensers

These condensers are provided at a high level with a long vertical discharge pipe as shown in Figure 1.21. In high level jet condensers, exhaust steam enters at the bottom, flows upwards and meets the down coming cooling water in the same way as that of low level jet condenser. The vacuum is created by the air pump, placed at the top of the condenser shell. The condensate and cooling water descends through a vertical pipe to the hot well without the aid of any pump. The surplus water from the hot well flows to the cooling water tank through an overflow pipe.

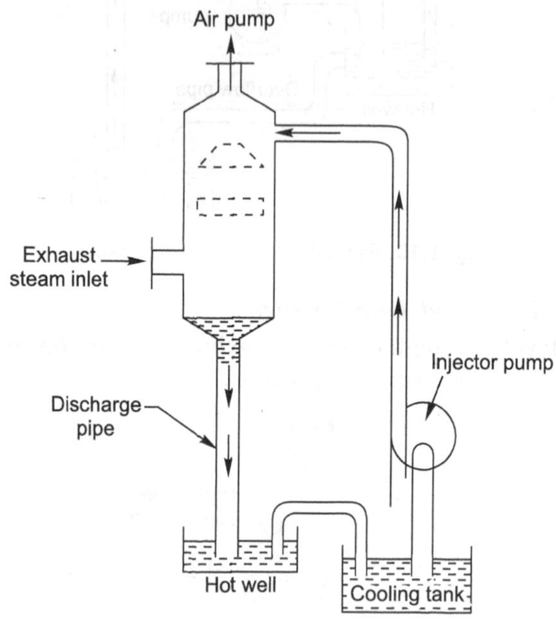

Fig. 1.21. High Level Jet Condensers

Ejector Condensers

In ejector condensers, the steam and water mix up while passing through a series of metal cones. Water enters at the top through a number of guide cones. The exhaust steam enters the condenser through non-return valve arrangement. The steam and air then passes through the hollow truncated cones. After that it is dragged into the diverging cones where its kinetic energy is partly transformed to pressure energy. The condensate and cooling water is then discharged to the hot well as shown in Figure 1.22.

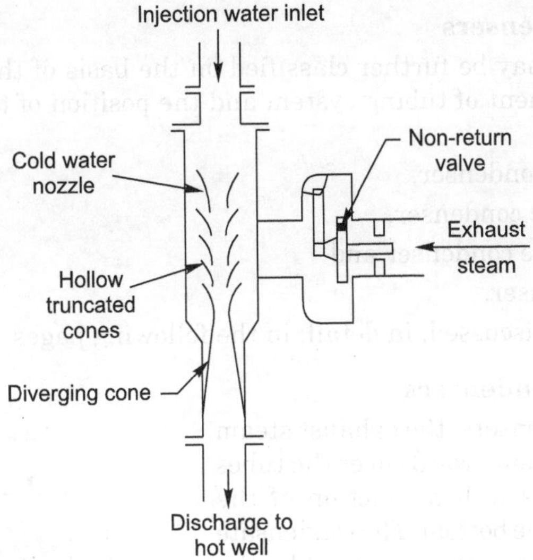

Fig. 1.22. Ejector Condenser

B. Surface Condensers

A surface condenser has a great advantage over the jet condensers, as the condensate does not mix up with the cooling water. As a result of this, whole condensate can be reused in the boiler. This type of condenser is essential in ships which can carry only a limited quantity of fresh water for the boilers. It is also widely used in land installations, where inferior water is available or the better quality of water for feed is to be used economically.

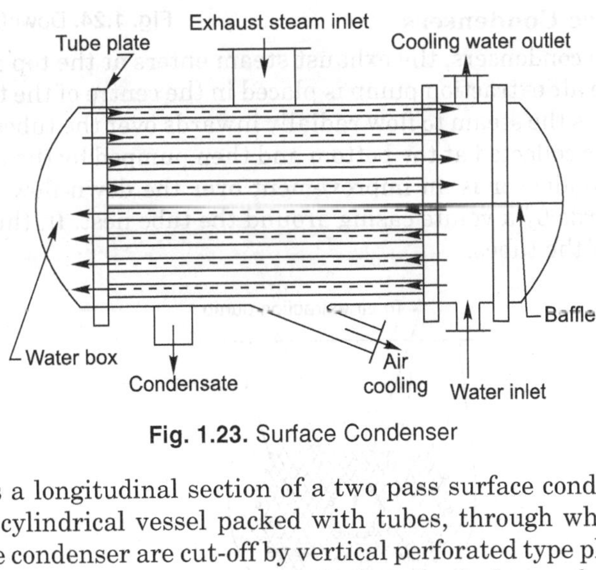

Fig. 1.23. Surface Condenser

Figure 1.23 shows a longitudinal section of a two pass surface condenser. It consists of a horizontal cast iron cylindrical vessel packed with tubes, through which the cooling water flows. The ends of the condenser are cut-off by vertical perforated type plates into which water tubes are fixed. This is done in such a manner that the leakage of water into the centre condensing space is prevented.

The water tubes pass horizontally through the main condensing space for the steam. The steam enters at the top and is forced to flow downwards over the tubes due to the suction of the extraction pump at the bottom.The cooling water flows in one direction through the lower half of the tubes and returns in opposite direction through the upper half, as shown in Figure 1.23.

Types of Surface Condensers

The surface condensers may be further classified on the basis of the direction of flow of the condensate, the arrangement of tubing system and the position of the extraction pump, into the following four types:

1. Down flow surface condenser,
2. Central flow surface condenser,
3. Regenerative surface condenser and
4. Evaporative Condenser.

These condensers are discussed, in detail, in the following pages.

Down Flow Surface Condensers

In down flow surface condensers, the exhaust steam enters at the top and flow downwards over the tubes due to force of gravity as well as suction of the extraction pump fitted at the bottom. The condensate is collected at the bottom and then pumped by the extraction pump. The dry air pump suction pipe, which is provided near the bottom, is covered by a baffle so as to prevent the entry of condensed steam into it, as shown in Figure 1.24.

As the steam flows perpendicular to the direction of flow of cooling water (inside the tubes), this is also called a cross surface condenser.

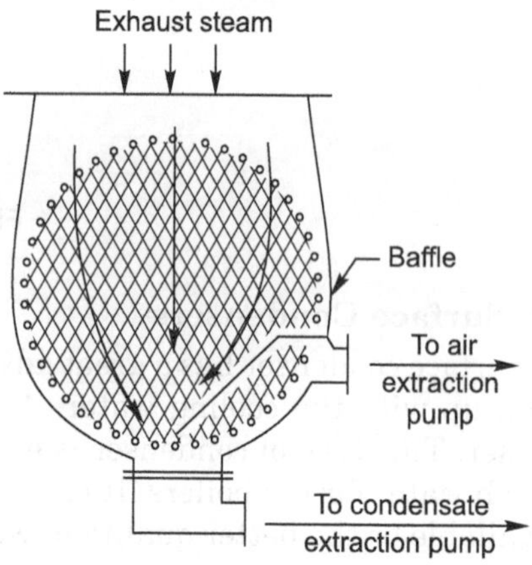

Fig. 1.24. Downflow Surface Condenser

Central Flow Surface Condensers

In central flow surface condensers, the exhaust steam enters at the top and flow downwards. The suction pipe of the air extraction pump is placed in the centre of the tube nest as shown in Figure 1.25. This causes the steam to flow radially inwards over the tubes towards the suction pipe. The condensate is collected at the bottom and then pumped by the extraction pump. The central flow surface condenser is an improvement over the down flow type as the steam is directed radially inwards by a volute casing around the tube nest. It, thus, gives an access to the whole periphery of the tubes.

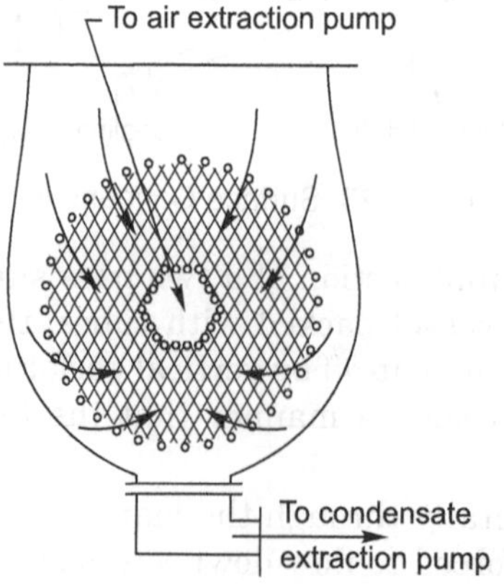

Fig. 1.25. Central Flow Surface Condensers

Regenerative Surface Condensers

In regenerative surface condenser, the condensate is heated by a regenerative method. The condensate after leaving the tubes is passes through the exhaust steam from the engine or turbine. It thus, raises its temperature for use as feed water for the boiler.

Evaporative Condenser

The steam to be condensed enters at the top of a series of pipes outside of which a film of cold water is falling. At the same time, a current of air circulates over the water film, causing rapid evaporation of some of the cooling water. As a result of this, the steam circulating inside the pipe is condensed. The remaining cooling water is collected at an increased temperature and is reused. Its original temperature is restored by the addition of the requisite quantity of cold water.

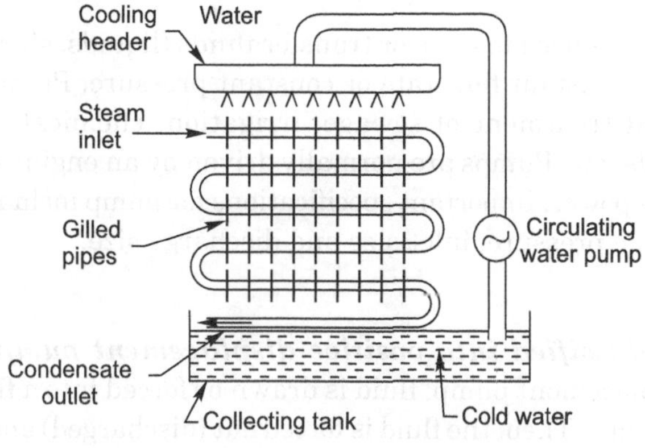

Fig. 1.26. Evaporative Condenser

The evaporative condensers are provided when the circulating water is to be used again and again. These condensers consist of sheets of gilled piping, which is bent backwards and forward and placed in a vertical plane, as shown in Figure 1.26.

3. Comparison of Jet and Surface Condensers

Following are the important points of comparison between jet and surface condensers:

S. No.	*Jet condensers*	*Surface condensers*
1.	Cooling water and steam are mixed up.	Cooling water and steam are not mixed up.
2.	Less suitable for high capacity plants.	More suitable for high capacity plants.
3.	It requires less quantity of circulating water. Condensate is wasted.	It requires a larges quantity of circulating water.
4.	The condensing plant is economical and simple.	Condensate is reused. The condensing plant is costly and complicated.
5.	Its maintenance cost is low.	Its maintenance cost is high.
6.	More power is required for air pump.	Less power is required for air pump.
7.	High power is required for water pumping.	Less power is required for water pumping.

Advantages of a Condenser in a Steam Power Plant

Following are the main advantages of incorporating a condenser in a steam power plant:

1. It increases expansion ratio of steam, and thus increases efficiency of the plant.
2. It reduces back pressure of the steam, and thus more work can be obtained.
3. It reduces temperature of the exhaust steam, and thus more work can be obtained.
4. The reuse of condensate (*i.e.*, condensed steam) as feed water for boilers reduces the cost of power generation.
5. The temperature of condensate is higher than that of fresh water. Therefore the amount of heat supplied per kg of steam is reduced.

1.6.4 Pumps

1. Introduction

A pump is a machine that is used to raise or transfer fluids (liquids, slurries and gases). Pump is also used to maintain constant flow rate or constant pressure. Pumps find applications in draining, collection and treatment of sewage, irrigation, chemical industries, petroleum industries, medical fields, etc. Pumps are normally driven by an engine or a motor. Pumps are commonly rated by horsepower. Important specifications for pump include maximum discharge flow, maximum discharge pressure, inlet size and discharge size.

2. Types of Pumps

Pumps are broadly classified into positive displacement pumps and rotodynamic pumps. In positive displacement pump, fluid is drawn or forced into a finite space and is then sealed by mechanical means. Then, the fluid is forced out (discharged) and the cycle is repeated. A positive displacement pump is one in which a definite volume of liquid is delivered for each cycle of pump operation regardless of the head against which the pump is operating and the resistance to flow offered by the system.

Reciprocating pump is a positive displacement pump. Some positive displacement pumps like gear pump and vane pump use rotary action instead of reciprocating action. In rotodynamic pumps, there is free passage of fluid between the inlet and outlet without any intermittent sealing. *Centrifugal pump is a rotodynamic pump.*

Priming: Before starting the centrifugal pump/reciprocating pump, priming is to be performed. Priming means filling the suction pipe and the casing with water in the case of centrifugal pump—suction pipe and clearance volume of the cylinder in the case of reciprocating pump, so that there is no air pocket left in the pump.

3. Reciprocating Pump

Reciprocating pump uses a piston and cylinder arrangement with suction and delivery valves integrated into the pump. Reciprocating pump can be single-acting (single suction and discharge strokes) or double-acting (suction and discharge in both the directions). There may be a single cylinder or multi-cylinder.

Uses of reciprocating pumps: *In general, reciprocating pumps are best suited for small capacities and high heads.* Reciprocating pumps were used extensively in steam power plant as boiler feed water pump. Reciprocating pumps are also frequently used in pneumatic and hydraulic systems for pumping highly viscous fluids including concrete and heavy oils. Use of reciprocating pumps in oil drilling operations is very common.

Advantages of Reciprocating Pumps

Reciprocating pumps are usually preferred due to their relatively compact design, high-viscosity performance and ability to handle high differential pressure.

Types of Reciprocating Pump

Reciprocating pumps are broadly classified as single-acting and double-acting.

4. Single Acting Reciprocating Pump

If the mechanical energy is converted into hydraulic energy (or pressure energy) by sucking the liquid into a cylinder in which a piston is reciprocating (moving backwards and forwards), which exerts the thrust on the liquid and increases its hydraulic energy (pressure energy), the pump is known as reciprocating pump. The main parts of a reciprocating pump are:

1. A cylinder with a piston, piston rod, connecting rod and a crank.
2. Suction pipe
3. Delivery pipe
4. Suction valve
5. Delivery valve

Working of a Reciprocating Pump

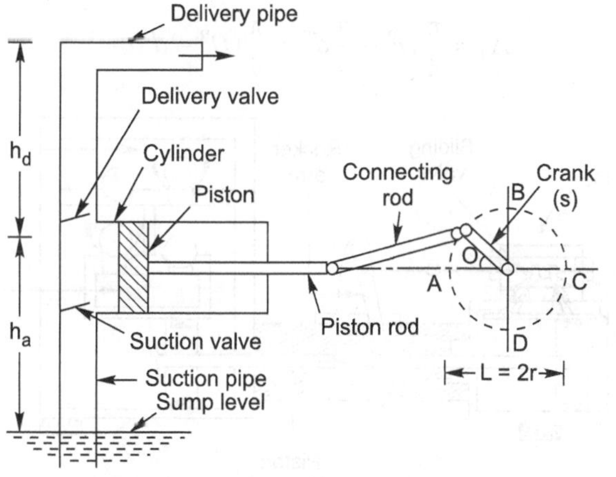

Fig. 1.27. Single acting Reciprocating Pump

The single acting reciprocating pump Figure 1.27, which consists of a piston which moves forwards and backwards in a close fitting cylinder. The movement of the piston is obtained by connecting the piston rod to crank by means of a connecting rod. The crank is rotated by means of an electric motor. Suction and delivery pipes with suction valve and delivery valve are connected to the cylinder. The suction and delivery valves are one way valves or non-return valves, which allow the water to flow in one direction only. Suction valve allows water from suction pipe to the cylinder which delivery valve allows water from cylinder to delivery pipe only.

When crank starts rotating, the piston moves to and fro inside a cylinder when crank is at A, the piston is at the extreme left position in the cylinder. As the crank is rotating from *A* to *C* (*i.e.*, from $\theta = 0$ to $\theta = 180°C$), the piston is moving towards right in the cylinder. The movement of the piston towards right creates a partial vacuum in the cylinder. But on the surface of the liquid in the sump atmospheric pressure is acting, which is more than the pressure inside the cylinder. Thus the liquid is forced in the suction pipe from the sump. This liquid opens the suction valve and enters the cylinder.

When the crank is rotating from C to A (*i.e.*, from $\theta = 180°$ to $\theta = 360°$), the piston moves from its extreme right position starts moving towards left in the cylinder. The movement of the piston towards left increases the pressure of the liquid inside the cylinder more than atmospheric pressure. Hence suction valve closes and delivery valve opens. The liquid is forced into the delivery pipe and is raised to a required height.

5. Double Acting Reciprocating Pump

In the double acting pump Figure 1.28, the water is acting on both sides of the piston. Thus we require two suction pipes and two delivery pipes for double-acting pump. When there is a suction stroke or one side of the piston, there is at the same time of a delivery stroke on the other side of the piston. Thus for one complete revolution of the crank there are two delivery strokes and water is delivered to the pipes by the pump during these two delivery strokes.

D = Diameter of the piston in m

d = Diameter of the piston rod or connecting rod in m

$\therefore$ Area on one side $A = \frac{\pi}{4}D^2$. Area on the other side of the piston, where piston rod is connected to the piston.

$$A_1 = \frac{\pi}{4}D^2 - \frac{\pi}{4}d^2 = \frac{\pi}{4}(D^2 - d^2)$$

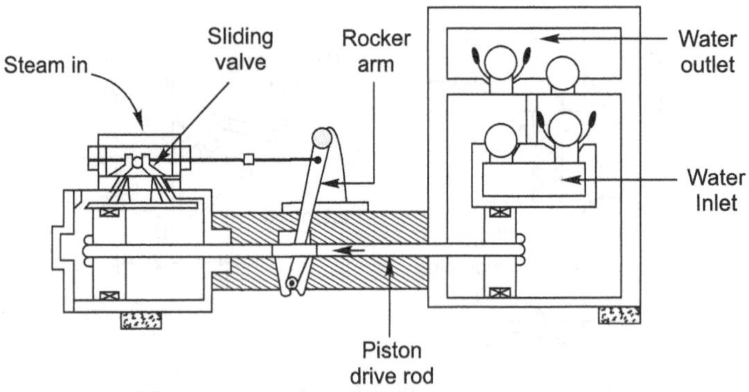

Fig. 1.28. Double Acting Reciprocating Pump

6. Centrifugal Pumps

- If the mechanical energy is converted into pressure energy by means of centrifugal force acting on the fluid, the hydraulic machine is called centrifugal pump Figure 1.29

Main parts of a centrifugal pump:

1. Impeller
2. Casing
3. Suction pipe with a foot valve and a strainer
4. Delivery pipe

Impeller: The rotating part of a centrifugal pump is called 'Impeller'. It consists of a series of backward curved varies. The impeller is mounted on a shaft which is connected to the shaft of an electric motor.

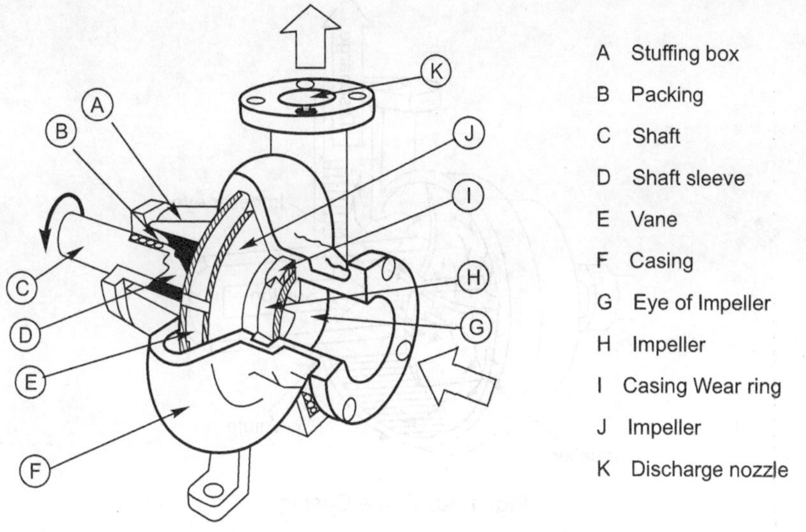

Fig. 1.29. Main Parts of the Centrifugal Pump

Casing: The casing of a centrifugal pump is similar to the casing of a reaction turbine. It is an airtight passage surrounding the impeller and is designed in such a way that the kinetic energy of the water discharged at the outlet of the impeller is converted into pressure energy before the water leaves the casing and enters the delivery pipe.

The following three types of the casings are commonly adopted:

(*a*) Volute casing
(*b*) Vortex casing
(*c*) Casing with guide blades

(*a*) **Volute casing:** It is of spiral type (Figure 1.30) in which area of flow increases gradually. The increase in area of flow decreases velocity of flow.

The decrease in velocity increases the pressure of the water flowing through the casing. It has been observed that in case of volute casing, the efficiency of the pump increases slightly as a large amount of energy is lost due to the formation of eddies in this type of casing.

(*b*) **Vortex casing:** If a circular chamber is introduced between the casing and the impeller. The casing is known as vortex casing Figure 1.31 (*a*). By introducing the circular chamber, the loss of energy due to the formation of eddies is reduced to a considerable extent. Thus the efficiency of the pump is more than the efficiency when only volute casing.

(*c*) **Casing with guide blades:** The impeller is surrounded by a series of guide blades mounted on a ring which is known as diffuser. The guide vanes are designed in which a way that the water from the impeller enters the guide vanes without stock. Also the area of the guide vanes increases, thus reducing the velocity of flow through guide vanes and consequently increasing the pressure of water. The water from the guide vanes then passes through the surrounding casing which is in most of the cases concentric with the impeller.

Suction Pipe with a Foot valve and a Strainer: A pipe whose one end is connected to the inlet of the pump and other end dips into water in a sump is known as suction pipe. A foot valve which is a non-return valve or oneway type of valve is fitted at the lower end of the suction pipe. The foot valve opens only in the upward direction. A strainer is also fitted at the lower end of the suction pipe.

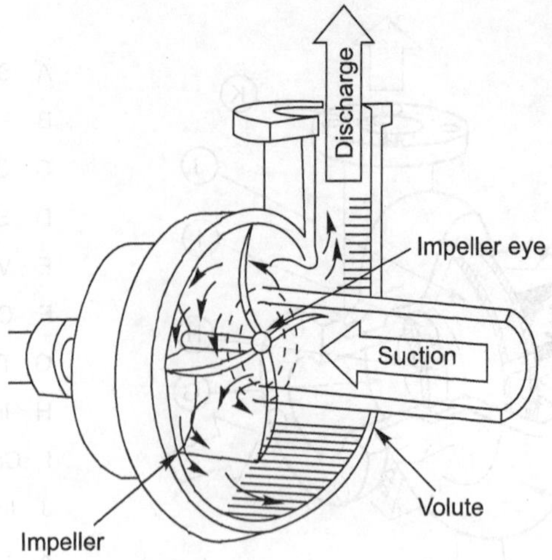

Fig. 1.30. Volute Casing

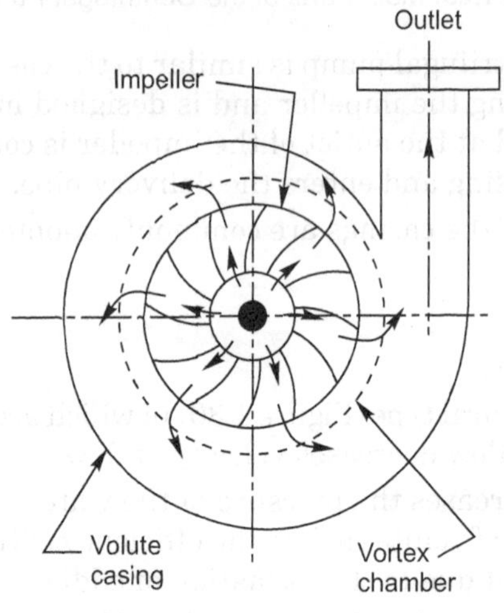

Fig.1.31 (a). Vortex Casing

Delivery pipe: A pipe whose one end is connected to the outlet of the pump and other end delivers the water at a required height is known as delivery pipe.

7. Types of Centrifugal Pumps

Centrifugal pumps are classified based on the direction of flow namely radial, mixed and axial flow with reference to the axis of rotation.

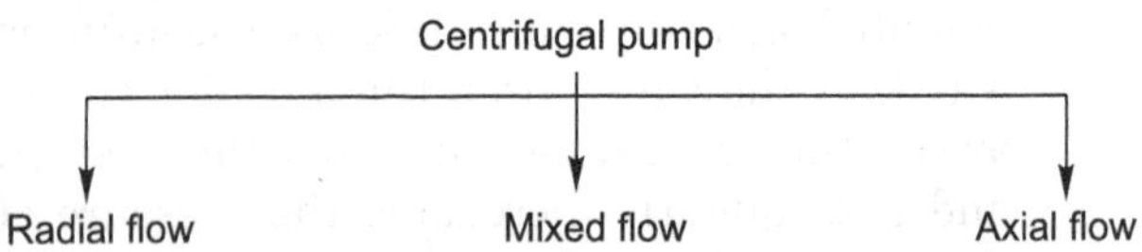

In the radial flow centrifugal pump, the pressure is developed wholly by centrifugal force. In mixed flow centrifugal pump, the pressure is developed partly by centrifugal force. In mixed flow centrifugal pump, the pressure is developed partly by centrifugal force and partly by the propelling action of the vanes of the impeller on the liquid. In axial flow centrifugal pump, the pressure is developed by the propelling action of the vanes of the impeller on the liquid.

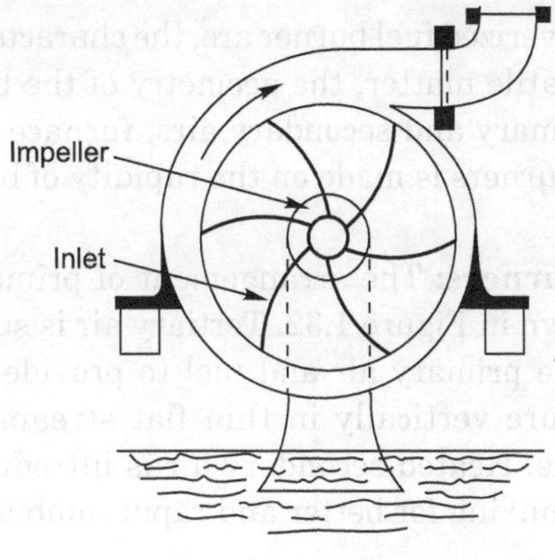

(*b*) Volute Centrifugal Pump

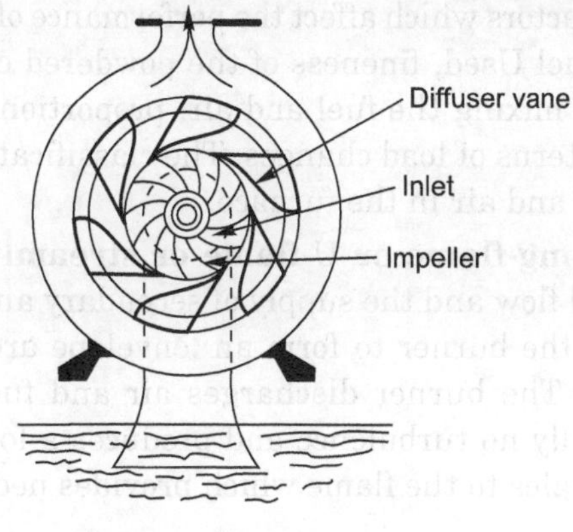

(*c*) B-Turbine Centrifugal Pump

Fig. 1.31. Types of centrifugal pump.

In volute centrifugal pump the area of flow gradually increases from the impeller outlet to the delivery pipe Figure 1.31 (*b*).

If a circular chamber is provided between the impeller and volute chamber the casing is known as Vortex Chamber Figure 1.31 (*c*).

8. Diffuser

- The impeller is surrounded by a diffuser.
- The guide vanes are designed in such a way that the water from the impeller enters the guide vanes without shock.
- It reduces the vibration of the pump.
- Diffuser casing, the diffuser and the outer casing are stationery parts.

1.7 PULVERISED COAL BURNERS

The efficient utilization of pulverized coal depends to a large extent upon the ability of the burners to produce uniform mixing of coal, air and turbulence within the furnace. The air which carries the pulverized fuel in the furnace through the burner is primary air and remaining secondary air required for complete combustion is admitted separately around the burner or elsewhere in the furnace. The pulverized coal burners should satisfy the following requirements:

(1) It should mix the coal and primary air thoroughly and project the same in the furnace properly with secondary air which is generally added around the burner.

(2) It should create proper turbulence and maintain stable combustion of coal and air throughout the operating range of the plant.

(3) It should control the flame shape and its travel in the furnace. This is generally done by the secondary air vanes and other control adjustments provided in the burner.

(4) The mixture of coal and air should move away from the burner at a rate equal to flame travel to avoid the flash back with the burner.

(5) The burner should also be provided with adequate projection against overheating, internal fire and excessive abrasive wear.

The factors which affect the performance of the pulverized fuel burner are, the characteristics of the fuel Used, fineness of the powdered coal, volatile matter, the geometry of the burner, place of mixing the fuel and air, proportions of primary and secondary airs, furnace design and patterns of load changes. The classification of burners is made on the rapidity of burning the coal and air in the furnace.

1. Long-flame or U-flame or streamlined burners: The arrangement of primary air and coal flow and the supply of secondary air is shown in Figure 1.32. Tertiary air is supplied around the burner to form an .envelope around the primary air and fuel to provide better mixing. The burner discharges air and fuel mixture vertically in thin flat streams with practically no turbulence and produces a long flame. Heated secondary air is introduced at right angles to the flame which provides necessary mixing for better and rapid combustion.

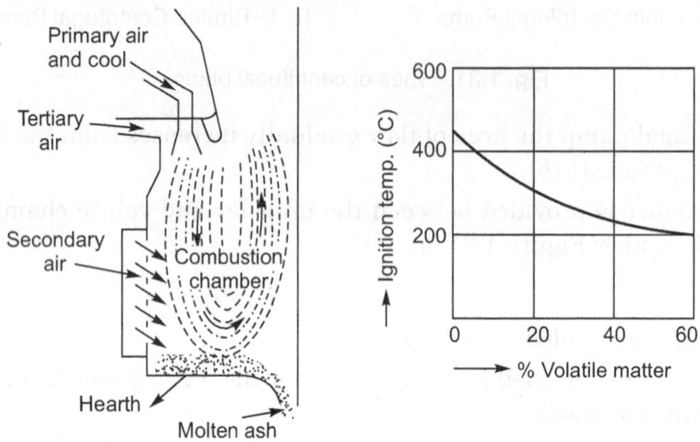

Fig. 1.32. Long-Flame or U-Flame or Streamlined Burners

Furnaces for low volatile coal are equipped with such burners to give a long flame path for slower burning of coal particles. The longer path provides more time to burn and its necessary to control the velocity in this zone (tip velocity is limited to 25 m/s). Less heat of ignition is available due to low volatile content and it is necessary to reduce the cooling effect from the wall tubes in the ignition zone by using a refractory belt round the furnace or by a refractory front wall. Generally low volatile coals have higher fusion temperature than bituminous coal and therefore higher furnace ratings are permissible.

2. Short flame or turbulent burner: The turbulent burners are usually set into furnace walls and project the flame horizontally into the furnace as shown in Figure 1.33. The fuel-air mixture and secondary hot air are arranged to pass through the burner in such a way that there is good mixing and the mixture is projected in highly turbulent form in the furnace. Due to high turbulence created before entering the furnace, the mixture burns intensely and combustion is completed in a short distance. This burner gives high rate of combustion compared with other types. The velocity at the burner tip is as high as 50 m/sec. The bituminous coal is successfully used with this burner. By proper adjustments, a long penetrating flame or short intensely hot flame can be produced. All modern plants use this type of burner. This is generally preferred for high volatile coals.

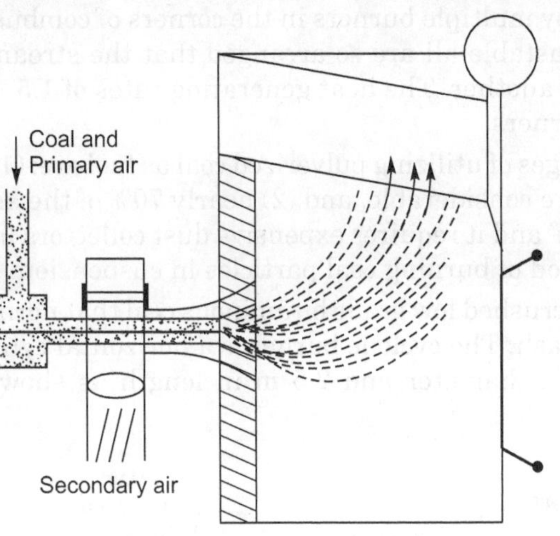

Fig. 1.33. Short flame or turbulent burner

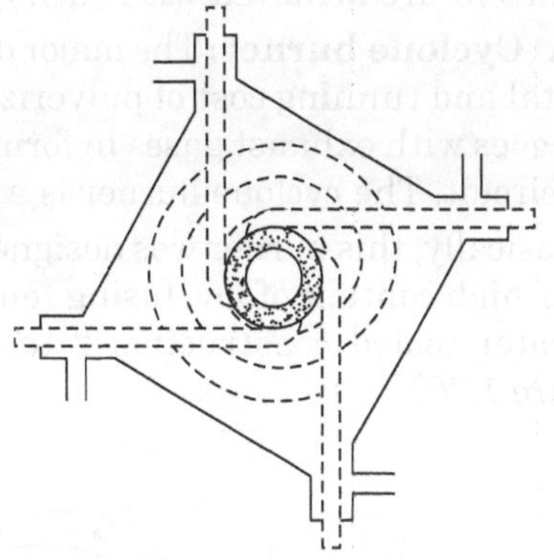

Fig. 1.34. Plan of furnace tangential firing

3. Tangential burners: The tangential burners are set in the furnace as shown in Figure 1.34 and discharge the fuel-air mixture tangentially to an imaginary circle in the centre of the furnace. The swirling action produces sufficient turbulence in the furnace to complete the combustion in a short period and avoids necessity of producing high turbulence at the burner itself. High heat release rates are possible with this method of firing.

This type of burner is sometimes constructed with tips that can be angled through a small vertical arc (± 30) so as to raise or lower the position 'of the turbulent combustion region in the furnace. This arrangement controls the temperature of the gases at the furnace aperture and maintains constant superheat temperature of the steam as the load varies: When the burners are tilted downward, the furnace gets filled completely with the flame and the furnace exit gas temperature is reduced as the furnace absorption is greater. This reduces the heat given to the super heater. The reverse is also true when the burners are tilted upwards. The usual limit of tilt ± 30° is sufficient to provide 100° C difference in the furnace gas exit temperature.

In a pulverized fuel firing system, the fluid ash carried in the gas stream is likely to be deposited on exposed metal surfaces and solidify thereon as slag. Sufficient heat absorbing surface must be provided to cool the molten ash by radiation below its softening temperature before it comes into contact with metallic surfaces. Thus, in the large units, a tall furnace is used and in very large units, it becomes economical to subdivide the furnace either by water-cooled division walls or by arranging side-by-side separately cased furnaces.

One of the main problems in pulverized fuel firing is to shorten the time required for combustion. The development of combustion chamber design has been marked throughout its history by the endeavor to burn the maximum amount of coal dust with minimum space and in minimum time.

For the given fineness of coal, shortening of combustion time is mainly the aerodynamic problem of bringing of air and fuel together within the furnace as quickly as possible. This means that air and fuel particles must be in violent motion (turbulence) relative to each other; Early designers sought to achieve—this, through *the use of burners* causing a turbulent or whirling motion, skillfully introducing primary, secondary and tertiary air. Such burners promote quick ignition but it is a characteristic of pulverized fuel firing that it is extremely difficult to influence the tail of combustion. This cannot be done by burner agitation or by introducing secondary air at point farther along in the path of the burning dust. The intense motion needed is secured in modern practice by multiple burners in the corners of combustion chamber or by telling burners which are adjustable all are so arranged that the streams of burning fuel and air impinge violently on one another. The heat generating rates of 1.5×10^6 $kJ/m^3/hr$ are achieved with such types of burners.

4. Cyclone burner: The major disadvantages of utilizing pulverized coal as fuel are: (1) the capital and running cost of pulverized mills are considerable, and (2) nearly 70% of the ash in coal goes with exhaust gases in form of 'fly-ash' and it requires expensive dust collectors in the gas circuit. The cyclone burner is a new method of burning coal particles in suspension.

Basically, this burner was designed to burn crushed low grade bituminous coal that normally has a high content of low fusing temperature ash. The cyclone burner is a horizontal cylinder of water cooled construction 2 to 3 meters in diameter and 2.5 m in length as shown in Figure 1.35.

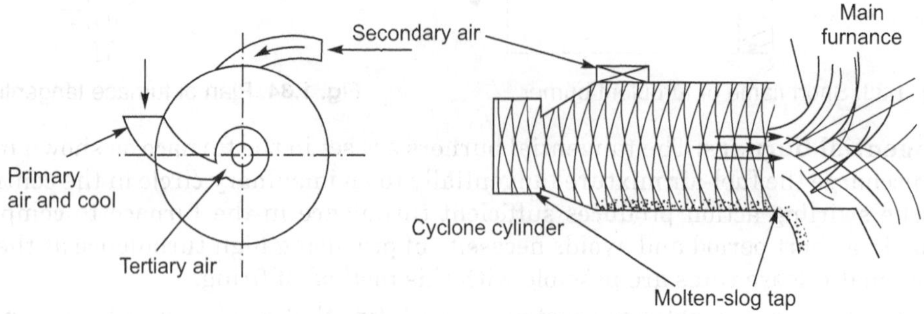

Fig. 1.35. Cyclone Burner

The inside part of the cyclone cylinder is lined with chrome ore. The horizontal axis of the burner is slightly deflected towards the boiler. It is 'externally' arranged to the boiler furnace and equipped with a single scroll type inlet at one end and a gas discharge throat in the boiler at the other end.

The coal used in cyclone burner is crushed to 6 mm maximum size and blown into a cylindrical 'cyclone furnace'. Air at 80 cm water pressure and coal admitted tangentially to the cylinder at outer end creates strong and highly turbulent vortex. As the coal with air moves from the front to rear, secondary air is introduced tangentially as shown in figure to complete the combustion. Extremely high heat liberation rate and the use of preheated air cause high temperature (2000°C) in the cyclone. The fuel supplied is quickly consumed and liberated ash forms a molten film flowing over the inner wall of the cylinder. The molten ash flows to an appropriate disposal system as the horizontal axis of the burner is tilted.

The merits of the cyclone burner over the other types are listed below:

1. Simplified coal crushing equipments can be used instead of costly pulverized mills.

2. All the incombustibles are retained in cyclone burner therefore the boiler fouling and all problems associated with it are reduced. A comparison between cyclone burners. and pulverized coal-firing equipment on the basis of 75 MW capacity plant shows 25% reduction in floor area; 33% reduction in building volume, 25% reduction in weight and 80% reduction in maintenance charges. This is because of the high furnace heat release rates.

3. As the forced draught is used with this type of burner, it can be operated with less excess air.Excess air required can be reduced to 15% minimum using forced draught fan.

4. Combustion rate can be controlled by simultaneous manual adjustment of fuel feed and air flow and response in firing rate changes is comparable to that of pulverized coal firing.

5. It is seen from the performance of the burner that the slag recovery is 82% and dust passing to the stack is 8% whereas in pulverized system, the percentages of slag and dust formation are exactly opposite. Therefore there is a considerable simplification of equipment required for the removal of dust from flue gases.

6. The cyclone furnace can use low grade fuels, reduces the size of steam generator and limits the fly-ash emission so that excessive furnace cleaning and precipitators are not required.

The heat release rates of the order of 16×10^6 to 20×10^6 kJ/m^3/hr and coals with high ash melting temperatures are successfully used in this type of burner.

The cyclone furnace gives best results with low grade fuels and high silica ratio (80%). Majority of the Indian coals can be straight away burned in cyclone furnace very successfully. A few percentage of Indian coals are such that can be used in cyclone furnace directly but it is possible to use with addition of limestone to reduce silica ratio from 97 to 84 and fusion temperature from 1400°C to 1300°C.

1.8 PRINCIPLE OF FLUIDISED BED COMBUSTION (FBC) SYSTEM

When a gas is passed through a packed bed of finely divided solid particles, it experiences, a pressure drop across the bed. At low gas velocities, this pressure drop is small and does not disturb the particles. But, if the gas velocity is increased further, a stage is reached when the particles are suspended in the gas stream and the packed bed becomes a fluidized bed. With further increase-in gas velocity, the bed becomes turbulent and rapidmixing of particles occurs. In general, the behaviour of this mixture of solid particles and gas is like a fluid. Burning of a fuel in such a state is known as fluidized bed combustion.

The arrangement of the FBC system is shown in Figure 1.36. The fuel and inert material dolomite are fed on a distributor plate and air is supplied from the bottom of the distributor plate. High velocity of air keeps the solid feed material in suspended condition during burning. The generated heat is rapidly transferred to the water passing through the tubes immersed in the bed and generated steam is taken out. During the burning, SO_2 formed is absorbed by the dolomite and prevents its escape with the exhaust gases. The molten slag is tapped from the surface of the bed.

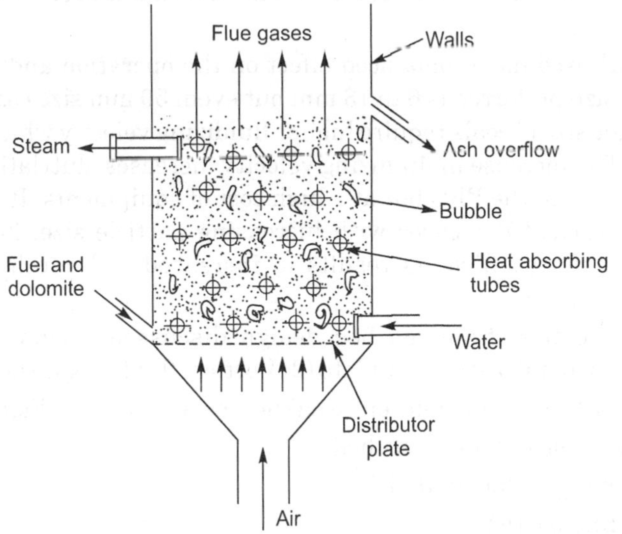

Fig. 1.36. Fluidized Bed Combustion System

The inert material is used with a primary object of controlling bed temperature and it accounts for 90% of the bed volume. The heat released by the combustion is first used in keeping up the temperature of the inert material and the balance is absorbed by the heat transfer surfaces. It is essential to choose the inert material judiciously as it remains with the fuel in continuous motion and at high temperature (800°C).The inert material should be resistant to heat and disintegration and should have similar density as that of coal. In addition to this, it should not disintegrate coal, the parent material of the bed. Sintered ash, fused alumina, sand, mullite, and zirconia are few suggested inert materials for FBC. However, the proper choice depends on the physical characteristics of coal. The amount of coal is only a small percentage of bed material (2% only). Since combustion tends to be almost complete while in intimate contact with hot lime stone.

The bed operates at much lower temperature (900°C) but still combustion efficiency remains very high (99.5%) as very high heat transfer rates are maintained over the surface of the tubes. Also the bed coal content is hardly 2 to 3% of bed volume. Thus, the presence of even a large amount of ash in coal is no disadvantage. Even the poorest grade coal could be burnt without sacrificing combustion efficiency.

The heat transfer rate to the tube surface is quite high as the system behaves like a violently boiling liquid and nearly 50% of the total heat released in the bed is absorbed by the tubes immersed in the bed leading to considerable reduction in the weight of the material and size of the boiler. The low and uniform operating temperature of the bed helps to prevent fusion of coal ash and therefore ash produced is soft and less abrasive.

Volatilization of ash constitutes is minimized and corrosion and erosion of submerged tubes are minimized as compared to conventional methods of combustion.

The bed operating temperature of 800–900°C is ideal for sulphur retention. Addition of limestone or dolomite to the bed brings down SO_2 emission level to 15% of that in conventional firing methods. Low NO_x emission is automatically achieved in FBC both due to low bed temperature and low excess air compared to a pulverized fuel furnace.

The cost economics shows that a saving of about 10% in operating cost and 15% in the capital cost could be achieved for a unit rating of 120 MW and it may be still higher for bigger units.

The size of the coal used has pronounced effect on the operation and performance of FBC system. The particle size preferred is 6 to 13 mm but even. 50 mm size coal can also be used in this system. But larger size of coals require higher fluidizing velocity which increases pressure drop across the bed. The increase in fluidizing velocity increases elutriation of carbon thereby increasing carbon back into the FBC but at a high cost of equipments. It is also observed that the heat transfer coefficient decreases with increasing particle size. For a given duty, the decrease in heat transfer coefficients results in increased bed depth to accommodate an increased heat transfer area.

The conventional vertical fluidized bed design discussed previously has three main characteristics: (1) Horizontal distributer plate (2) Vertical fluidizing airflow and vertical walls.

The conventional FBC has four inherent weaknesses as shown in Figure 1.37.

(1) Short residence time of fuel in the bed.

(2) Poor lateral mixing of 'fuel and air'.

(3) Elutriation of fine particles.

(4) Ash and heavy incombustibles are difficult to remove.

1. Short Residence Time

The repercussions are listed below:

(*a*) Carbon burn out is not optimum.

(*b*) Reduction of toxic products of combustion is not optimum.

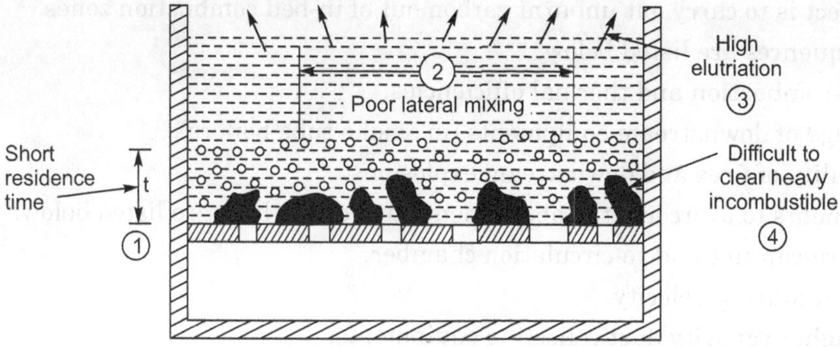

Fig. 1.37. Inherent weaknesses of FBC

The consequences are listed below:

(*a*) Lower combustion and thermal efficiency and lower output.

(*b*) Incomplete combustion of toxic gases.

(*c*) More equipments to recapture and recirculate unburned carbon.

(*d*) More secondary air to oxidize CO to CO_2

(*e*) More sorbent to kill SO_x

The treatments to overcome the above-mentioned difficulties are listed below:

(*a*) The feed stock should be fine.

(*b*) Lower the bed velocity.

(*c*) Recirculate the unburnt carbon.

(*d*) Supply more sorbent than optimum.

(*e*) Avoid feeding light solids and liquids.

2. Poor Horizontal Mixing

The repercussions are listed below:

(*a*) Bed fails to mix thoroughly fuel, sorbet and air.

The consequences are listed below:

(*a*) Part of bed is starved of fuel.

(*b*) Other parts of the bed are clinkered.

(*c*) Non-uniform temperature causes problem in control system.

(*d*) Heavy solids sink to bottom and locally block the bed.

(*e*) It creates difficulty in starting the bed from cold.

The treatments to overcome the above-mentioned difficulties are listed below:

(*a*) The fuel should be distributed with more feed points and splitters.

(*b*) Bubble caps are to be introduced to prevent the blockage.

(*c*) Use higher than optimum level of excess air.

(*d*) Use higher than optimum Ca : S ratio.

(*e*) Start up on shallow bed and deepen the bed gradually as temperature rises.

(*f*) Avoid using smoky and smelly fuels as this causes clinkering of the bed.

3. Elutriation of Fine Particles

The main effect is to carry out unburnt carbon out of in-bed combustion zones.

The consequences are listed below:

(*a*) Lower-combustion and thermal efficiencies.

(*b*) Blockage of downstream equipments (in smoke tube boilers).

(*c*) Possibility of fires and downstream explosions.

The treatments to overcome the above-mentioned difficulties are listed below:

(*a*) Add a re-capture and re-circulation chamber.

(*b*) Lower fluidizing velocity.

(*c*) Add higher capacity dust collecting equipments.

(*d*) Increase bed material particle size.

(*e*) Avoid feeding light particles.

(*f*) Add a cyclone in the furnace zone.

4. Ash Removal Difficulty

The main repercussions are listed below:

Heavy incombustibles block the distributor and destroy the fluidized bed.

The consequences are listed below:

(*a*) Loss of full fluidization.

(*b*) Less air distribution hampers proper bubbling.

(*c*) Reduced mixing and therefore thermal output is reduced.

(*d*) Creates cold and hot spots

(*e*) Downtime is more.

The treatments to overcome the above-mentioned difficulties are listed below:

(*a*) Introduce ash slots.

(*b*) Introduce bubble caps.

(*c*) Provide a defluidized layer at bottom of the bed.

(*d*) Blow-off ash.

Types of FBC

The FBCs are mainly classified into three basic systems which are outlined as given below:

1. Atmospheric Fluidized Bed Combustor. It is most widely applied and fully developed system at present. The commonly used underfeed and overfeed systems are shown in Figures 1.38 (*a*) and 1.38 (*b*) Underfeed system provides positive load and a compact design but costly in operation. Overfeed system is simple in operation and economical in running but results in smaller output per m^2 area and gives poor desulphurization performance. In this system, the pressure inside the bed is atmospheric.

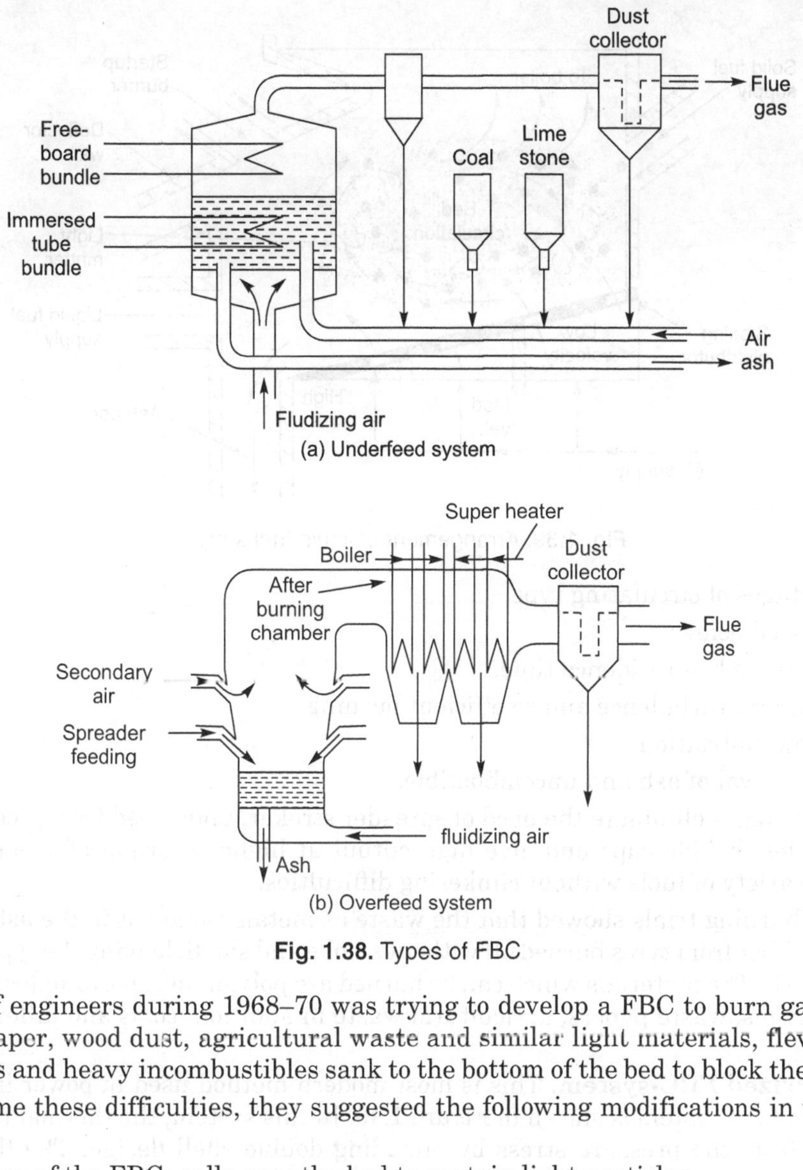

Fig. 1.38. Types of FBC

A group of engineers during 1968–70 was trying to develop a FBC to burn garbage. They found that paper, wood dust, agricultural waste and similar light materials, flew off the bed whilst metals and heavy incombustibles sank to the bottom of the bed to block the distributor.

To overcome these difficulties, they suggested the following modifications in the previous FBC :

(*a*) Bent one of the FBC walls over the bed to contain light particles.

(*b*) Provide sloping distributor plate to give an air-slide effect.

(*c*) Provide non-uniform fluidizing velocities over the bed.

These suggested modifications showed significant improvement and allowed to use light materials as fuels most successfully. Light materials were burned within the bed and heavy incombustibles. (ash and metals) gathered at the bottom of the slopping distributor.

The other outstanding features of this arrangement are ash pot at lower end of the distributor plate, secondary air supply over the bed, start up burner, and arrangement of liquid fuel supply as shown in Figure 1.39.

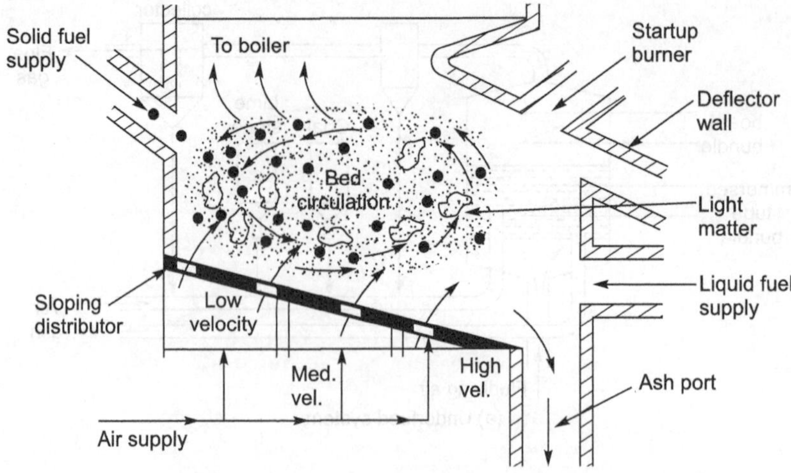

Fig. 1.39. Arrangement of liquid fuel supply

The advantages of circulating type FBC are listed below:

(*a*) Increased in bed residence time.

(*b*) High lateral turbulence and so efficient burning.

(*c*) Very low elutriation.

(*d*) Easy removal of ash and uncombustible.

These advantages eliminate the need of spreader stroker, under-bed feeds, recirculation of carbon particles, bubble caps and give high output at higher thermal efficiency. This also helps to use variety of fuels without clinkering difficulties.

Extensive burning trials showed that the waste as metals came out in the ash, bright and clean while rubber from tyres burned off without smoke and smell, leaving the type reinforcing metal in the ash. The materials which can be burned are polyamides, polyamides, tyre rubber, PVC, wax cartons, waste papers, agricultural waste in addition, coals and different forms of oils.

2. Pressurized FBC-system. This is most modern method used in power industry. The arrangement of the system is shown in Figure 1.40. In this system, the thermal reactor stress is separated from the pressure stress by providing double shell design. The flue gas after passing through dust collector is passed through a gas turbine which can run the compressor to supply pressurized air to FBC for fluidization and combustion. The advantages of pressurized system over atmospheric system are listed below:

* Substantial increase of coal loading (kg/m^2) with simultaneous reduction of air velocity.
* Considerable volume reduction of the system as shown in Figure 1.40.
* High burning rates.
* Improved desulphurization and low NO_x emission.
* Considerable reduction in the cost.

But it is difficult to control and give long life to the plant.

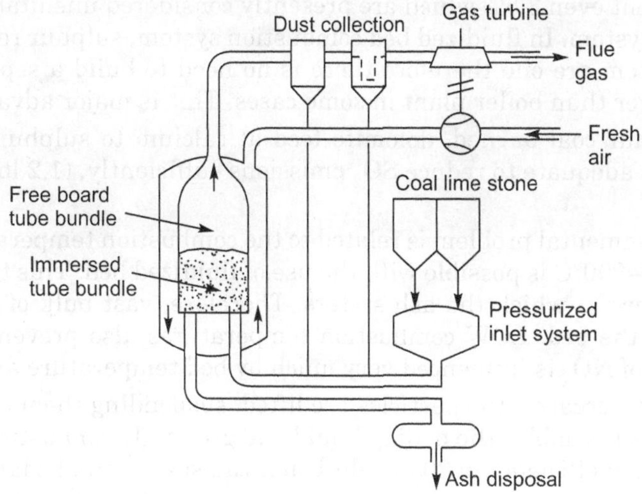

Fig. 1.40. Pressurized FBC-system

Advantages of Fluidized BED Systems

(1) The importance of the fluidized bed lies in the fact that it can use solid, liquid or gaseous fuel or mix as well as domestic and industrial waste. Any variety of coal irrespective of rank, ash content, ash fusion temperature, sulphur content can be made use of successfully. Major advantage is the capability of switching from one type of fuel to another, enabling the operator to take advantage of the cheapest fuel available on day-to-day basis. Virtually any type of combustible matter can be burned by adjusting the factors as size, air velocity and rate of feed.

(2) High combustion intensity (kJ/m^2-hr) can be achieved and varied with variation in air pressure supplied to the combustor. A heat rate 3×10^3 kJ/m^2-sec can be achieved even at atmospheric pressure and a fluidizing velocity of about 3 m/sec. This means that the combustor is much smaller than conventional furnaces. Small combustion chamber cuts the construction time required in addition to its capital cost with which its saves 15 to 20% towards its construction cost. Therefore, it is f:hyaper in capital and running cost.

(3) Solid mixing is extremely rapid and therefore high heat transfer rates can be obtained to surfaces immersed in the bed, plus a more effective use of tube surface owing to its immersion within the bed. This can lead to a saving of 75% in tube requirements.

(4) Combustion temperature can be controlled accurately and it can be low enough (750°C – 900°C) to minimize volatization of ash constituents like alkali metals (that can form deposits upon condensation and can cause corrosion problems in conventional boilers) because the temperature is well below the melting point of most gas-borne solid particles. Also, because ash particles have not been melted, they are soft and non-abrasive, unlike those from previous attempts at solid fuel gas turbine firing, where hard glassy particles erode the turbine blading.

(5) The system can readily be designed for operation at raised combustion pressure, owing to the simplicity of arrangement, small size of the plant and reduced likelihood of corrosion or erosion of gas turbine blades.

(6) The SO_2 formed due to combustion of sulphur can be absorbed in the fluidized bed if some of the mineral particles are $CaCO_3$. A chemical reaction occurs between SO_2 and $CaCO_3$ to form $CaSO_4$. The $CaSO_4$ being solid can be collected with the ash. Thus, we can use coals of

higher sulphur content even 3.5% which are presently considered unsuitable without the use of flue gas cleaning system. In fluidized bed combustion system, sulphur retention equipment and combustion system are one therefore there is no need to build a separate gas cleaning system which is bigger than boiler plant in some cases. This is major advantage of FBC.

(7) With 2% sulphur coal burned, dolomite feed at calcium to sulphur mole ratio in the range of 0.5 to 1 was adequate to reduce SO_2 emissions sufficiently, (1.2 lb of SO_2 per million BTU heat input).

(8) Another environmental problem is related to the combustion temperature. Stabilization of combustion at 700–900°C is possible with the use of fluidized bed. This temperature is well below the temperature at which the ash sinters. Therefore, vast bulk of ash can be tapped from tlle bottom of the bed. LOW combustion temperatures also prevent the formation of NO_x. The formation of NO_x is influenced very much by bed temperature and excess air.

(9) As gas velocity increases, the particles are lifted, suspending them in a turbulent mass and the floating mass resembles the boiling liquid and give high combustion efficiency 99.9%. One percent increase in efficiency of 30 tons/hr boiler can save around 2 lakh rupees per year.

(10) The combustion in conventional system becomes unstable when the ash exceeds 48% but even 70% ash containing coal can be efficiently burned in FBC. Flue gases from conventional oil-fired boiler plant contains some 300–500 ppm NO_x and from coal fired plant some 400 – 800 ppm. By contrast, from fluidized combustion of oil, the NO_x emission remained in the range of 100 ± 120 ppm and with coal under pressure, 120 ± 60 ppm. When burning propane the NO_x concentrations have been less than 40 ppm. This is comparable with the general limit imposed for US power stations of 700 ppm in ordinary areas and 100 ppm in densely populated areas.

(11) Pulverisation of solid fuel is not necessary with fluidized bed combuster. The maximum size of the coal particle is limited to 6 rom. Most of the research work is done using coal as a fuel as it is available in abundance.

(12) The volatilization of alkali compounds does not occur at bed temperature (800–900°C), so the deposition of the rubes is minimized. Low temperature during combustion reduces the fire side fouling.

(13) The large quantity of bed material acts as a thermal storage which reduces the effect of any fluctuations in fuel feed ratio.

The fluidized bed combustion system is particularly more attractive with high ash fuels (tar sands oil shales, coal refinery slurry) which can be burned at ease and where there is strict atmospheric pollution legislation.

1.9 COAL HANDLING SYSTEMS

In large power stations, it is not possible to handle large quantities of coal manually, therefore, some mechanical handling system must be introduced to the plant for easy and smooth operation and better control.

The coal handling starts at the discharge end of the carrier system (rail head or harbor) and passes through various stages such as delivery to storage, reclamation and feeding to the combustion system.

The implant coal handling system should be designed in such a way that the implant transportation should be minimum. The following points should be kept in mind while desiglling the coal handling plant:

1. The handling method should be simple and sound and require minimum operations and transport.

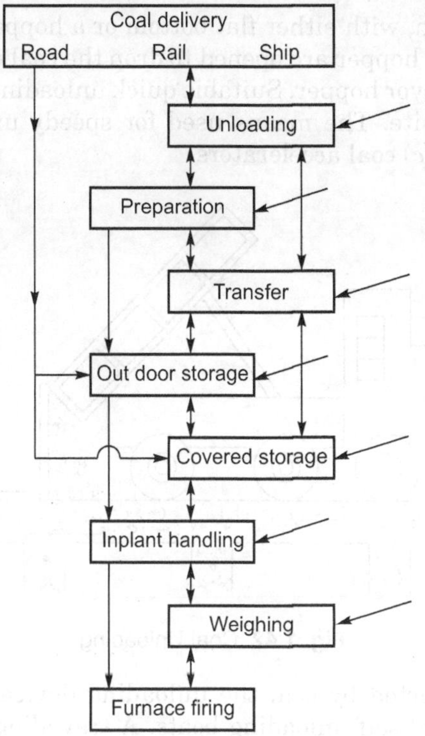

Fig. 1.41. Layout of Coal Handling System

2. There should not be double handling of coal in the plant.
3. The handling units should be centralized to facilitate inspection and maintenance.
4. The electrical motors should be used as prime-mover as much as possible as they are reliable, flexible and with high residual value.
5. The working parts should be enclosed to avoid abrasion and corrosion.
6. It should be able to deliver required quantity of coal during peak hours.

The coal handing plant capacity is generally three times of its normal requirement. It is not possible to give outline of a coal handling plant as its complexity depends on how coal is received, how the plant is situated and flexibility of the components required. The general flow of coal in the thermal power plant is shown in Figure 1.41. Typical devices required for each step of handling are also indicated on the flow chart. It is not necessary to follow the flow chart shown in Figure 1.41 for all plants. Some intermediate steps may be eliminated or rearranged depending upon the type of the plant, capacity of the plant and load factor.

Coal unloading and required equipments. The type of coal unloading equipment used in the plant depends upon the type of out-plant handling mode as road, rail or ship.

Road transportation of coal by trucks is often more economical for small and medium capacity plants located within 30 to 50 kilometers of coal mines. The use of lift trucks—with scoops is used for coal unloading as the entire load, of the truck can be easily unloaded without additional equipments. The arrangement of unloading as shown in Figure 1.42. The coal is dumped directly into unloading hoppers which delivers the coal to hoists or conveyors. If the coal is being stored in pile, the trucks travel up ramps and dump coal directly on the pile and bulldozers spread and compact the coal on pile. The railroad movement of coal is generally used for large coal quantities and for longer distances where water transportation is not available. The cars of 50 to 70 tons capacity each, with either flat bottom or a hopper bottom are used. In either case, the gates in the floor or hopper are opened to drop the coal out into an unloading hopper or directly onto a small conveyor hopper. Suitable quick unloading devices of coal cars must be provided at the unloading site. The means used for speedy unloading are (*a*) car-towers, (*b*) rotary tar dumpers, and (*c*) coal accelerators.

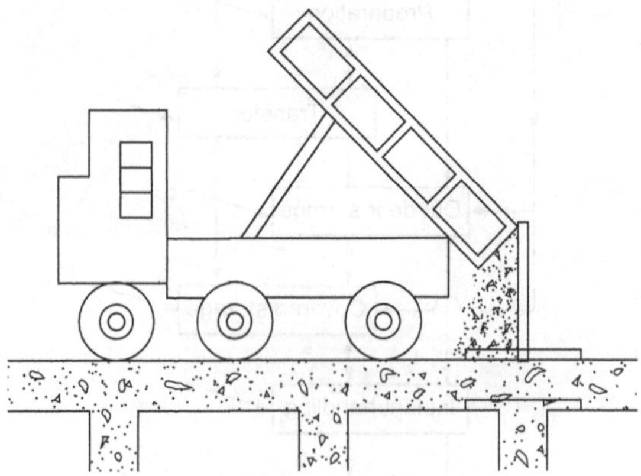

Fig. 1.42. Coal Unloading

When the coal is transported by sea, the unloading devices used are: (*a*) coal towers, (*b*) unloading bridges, and (*c*) self-unloading boats. A travelling gantry crane on unloading bridges is used for unloading large coal carrying vessels. Some modern boats are equipped with their own unloading conveyors and long, swinging belt conveyor boom which can be adjusted to carry the coal from the vessel to the desired storage point on the shore.

Coal preparation and required equipments. The preparation of coal before feeding to the combustion chamber is necessary if unsized coal is brought to the site. The coal preparation plant either can be located near the coal receiving point or at the point of actual use. The coal preparation plant includes the following equipments : (*a*) Crushers, (*b*) Sizers, (*c*) Dryers and (*d*) Magnetic separators.

The coal crushers are absolutely necessary to prepare the coal of required size before feeding to the combustion chamber part of the coal obtained from mines does not require sizing and is bypassed. The capacity of the coal Crushing plant must be sufficient to meet the peak load requirement. The capacity of crushers used in central plant is as high as 600 tons per hour. Sizers are used along with the crushers for the separation of coal of required size. The crushed coal is passed over the sizer which removes unsized coal and feeds back to the crusher. The sized coal is further passed to the drier to remove the moisture from the coal. The coal driers are used in order to remove the excess free moisture from the supplied coal or if it is wetted during transport. Hot flue gases are passed through the coal storage in closed spaces for removing the moisture from coal.

Before supplying the coal to the storage hopper, the iron scrap and particles are removed with the help of magnetic separators. The removal of the iron particles is necessary as they may choke the burners and may increase the wear of the handing equipment. The separator is usually a magnetized pulley over which the loaded coal belt is run. The iron particles cling to the belt as it travels around the magnetised pulley, whereas the coal falls off sooner. The iron particles drop off into a reject chute as the belt leaves the pulley. Thus, the coal and iron are quickly separated. The arrangement of the coal preparation plant is shown in Figure 1.43.

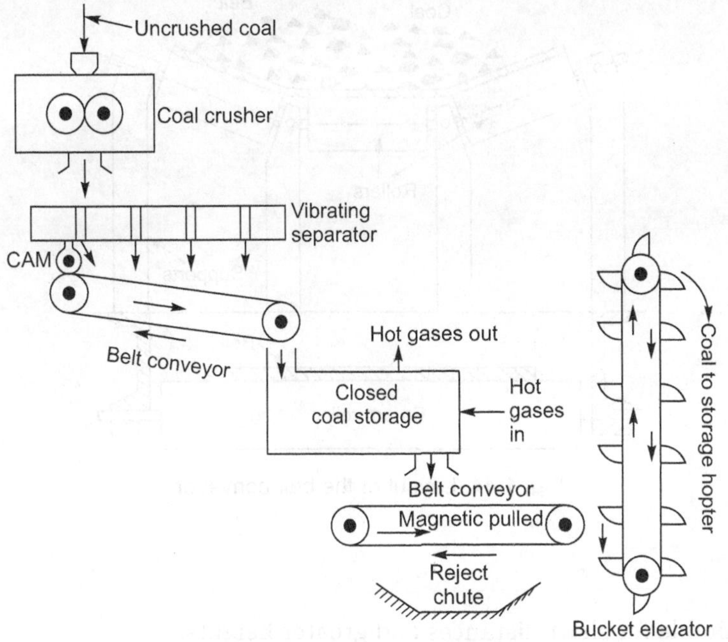

Fig. 1.43. Coal proparation plant

Transfer of Coal and Related Equipments. Transfer of coal includes the carrying of coal from unloading point to the storage site. It may require one equipment or several depending upon local condition. The type of equipments used for transferring the coal are listed below:

(*a*) Belt Conveyors, (*b*) Screw Conveyors,
(*c*) Bucket Elevators, (*d*) Grab Bucket Elevators,
(*e*) Skip Hoists, and (*f*) Flight Conveyors.

The details of each are discussed below:

(*a*) Belt Conveyors. This is very suitable means of transporting large quantities of coal over large distances. Belt conveyor consists of endless belt made of rubber, canvas *or* balata running over a pair of end drums *or* pulleys and supported by a series of rollers (known as idlers) provided at regular intervals. The return idlers which support the empty belt are plain rollers and are spaced wide apart. The layout of the belt conveyor is shown in Figure 1.44.

The initial cost of this coal carrying system is not high and power consumption is also low. The belt conveyors are successfully used on inclination up to 20° to the horizontal. An average speed of the belt conveyor is 60 m to 100 m per minute. The load carrying capacity of the belt may vary from 50 to 100 tonnes/hr and it can easily be transferred through 400 meters.

The advantages and disadvantages of belt-conveyor are listed below:

Advantages

1. This is most economical method of coal transport in medium as well as large capacity plants.
2. The rate of coal transfer can be easily varied by just varying the speed of the belt.
3. The repair and maintenance costs are minimum.
4. The coal over the belt can be easily protected from wind and rain just by providing overhead covers.
5. The power consumption to carry the coal is minimum compared with other conveyors.

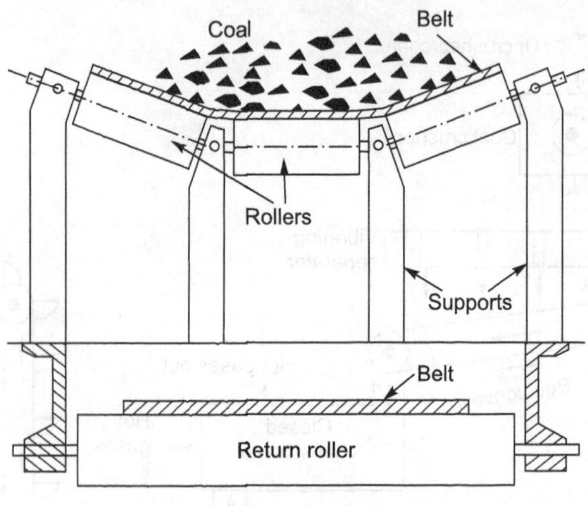

Fig. 1.44. Layout of the belt-conveyor

Disadvantages

1. It is not suitable for short distances and greater heights.
2. It cannot be used to carry the coal at greater heights as its inclination is limited to 20°.

***(b)* Screw Conveyor**. It consists of an endless helicoid screw fitted to a shaft as shown in Figure 1.45. The driving mechanism is connected to one end of the shaft and the other end of the shaft is supported in an enclosed ball bearing. The screw while rotating in a trough transfers coal from one end to the other end as shown in figure. The diameter of screw varies from 15 cm to 50 cm and its speed varies from 70 to 120 r.p.m. as per the capacity required. The maximum capacity of this conveyor is 125 tonnes per hour.

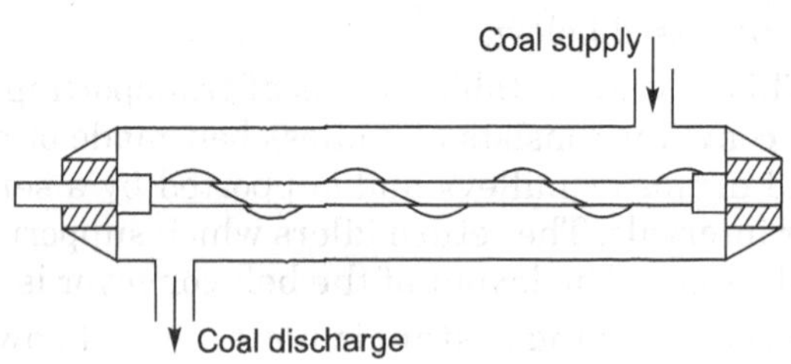

Fig. 1.45. Screw conveyor

The advantages and disadvantages of screw conveyor are listed below:

Advantages

1. It requires minimum space and is cheap in the first cost.
2. It is most simple and compact.
3. It can be made dust tight.

Disadvantages

1. The power consumption per unit weight transferred is considerably high.
2. The length of feed hardly exceeds 30 meters due to torsional strain on the shaft.

3. The wear and tear is very high and therefore the life of the conveyor is considerably short compared with belt-conveyor.

***(c)* Bucket Elevators.** This conveyor is extensively used for vertical lifts. It consists of buckets fixed to a chain which moves over two wheels as shown in Figure 1.46 (*a*). The coal is carried by the buckets from the bottom and discharged at the top. Another continuous type bucket elevator shown in Figure 1.46 (*b*) carries more amount of coal compared with first.

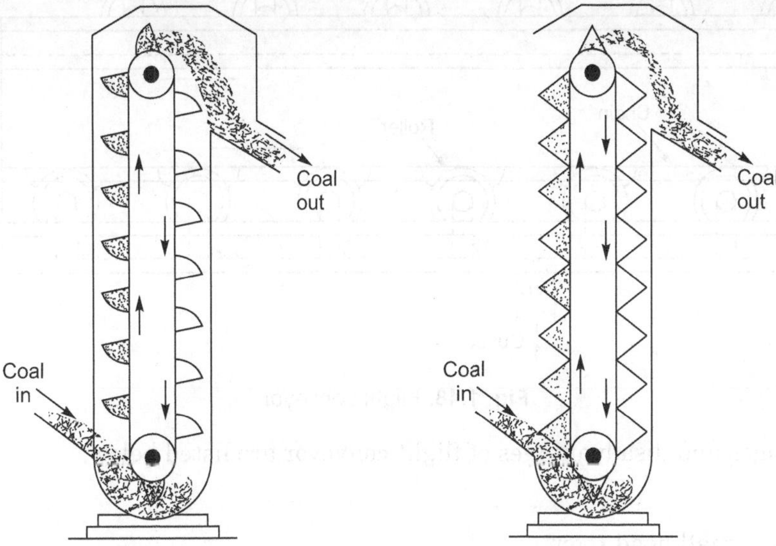

Fig. 1.46. Bucket elevators

The maximum height of the elevator is limited to 305 m (100 ft) and maximum inclination to the horizontal is limited to 60°. The speed of the chain required in first case is 75 m/min and continuous type is 35 m/min for about 60 tonnes capacity per hour.

***(d)* Grab Bucket Conveyor.** A grab bucket conveyor lifts as well as transfers the cooal from one point to another. The grab conveyor can be used with crane or tower as shown in Figure 1.47. A 2 to 3 cu-m bucket operating over a distance of 60 m transfers nearly 100 tonnes of coal per hour. The initial cost of this machine is high but operation cost is less. Its use for transferring the coal is justified only when other arrangements are not possible.

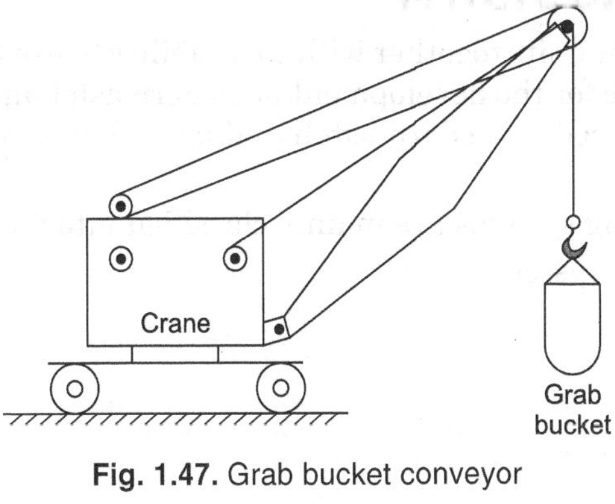

Fig. 1.47. Grab bucket conveyor

(*e*) **Flight Conveyor.** This conveyor is generally used for transfer of coal when filling of number of storage bins situated under the conveyor is required. It consists of one or two strands of chain, to which steel scrapers are attached. The scraper scraps the coal through a trough and the coal is discharged in the bottom of the trough as shown in Figure 1.48.

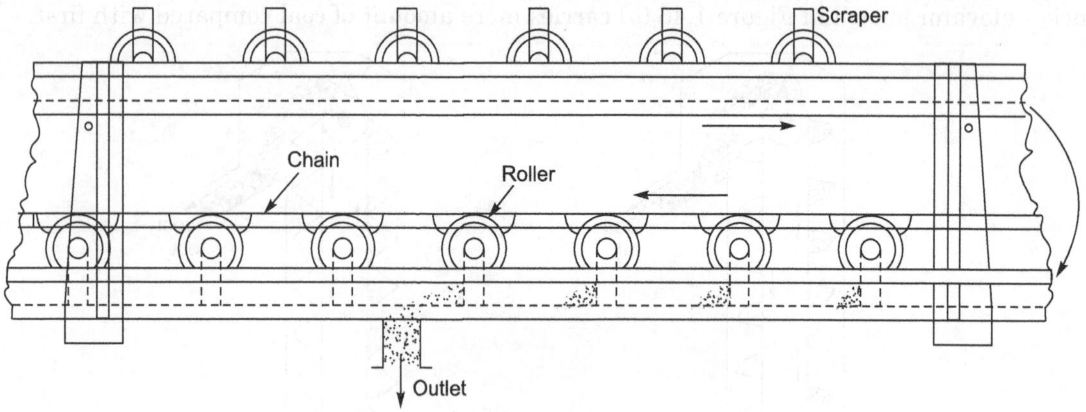

Fig. 1.48. Flight conveyor

The advantages and disadvantages of flight conveyor are listed below.

Advantages

1. It requires small head room.
2. The speed of the conveyor can be regulated easily to suit the requirements.
3. It can be used for coal as well as ash transfer.

Disadvantages

1. There is excessive wear and tear due to scrapping action and therefore the life is short.
2. The repair and maintenance charges are very high.
3. The speed is limited to 300/min. to reduce the abrasive action of the material handled.
4. Power consumption per unit of coal or ash transferred is considerably high due to dragging action.

1.10 ASH HANDLING SYSTEM

The ever increasing of boiler units together with their ability to use low grade high ash content coals have been responsible for the development of modern ash handling systems.The general layout of the components used in modern ash handling and dust collection plant is shown in Figure 1.49.

The modern ash-handling systems are mainly classified into four groups:

1. Mechanical handling system.
2. Hydraulic system.
3. Pneumatic system.
4. Steam jet system.

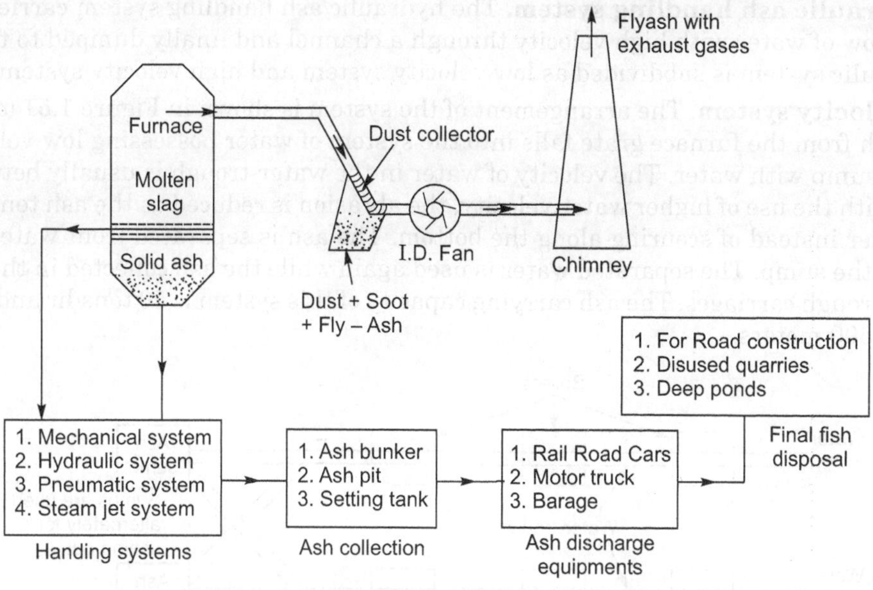

Fig. 1.49. General layout of ash handling and dust collection system

1. Mechanical ash handling system. The mechanical ash handling system is generally used for low capacity power plants using coal as fuel. The arrangement of the system is shown in Figure 1.50.

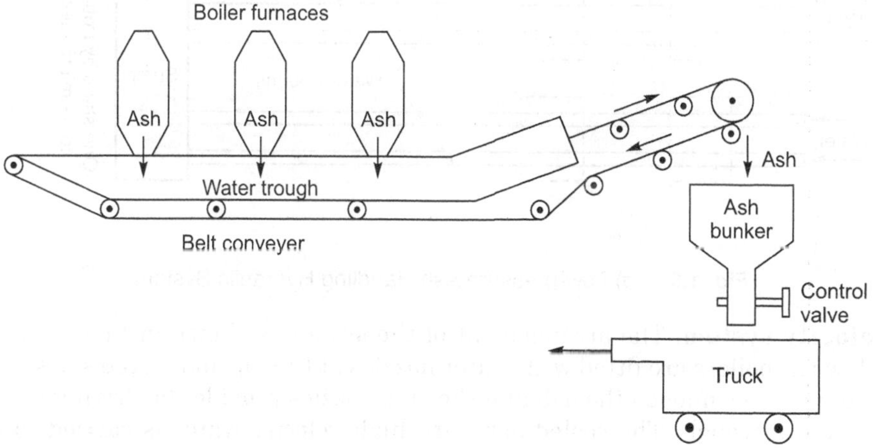

Fig. 1.50. Mechanical ash handling system

The hot ash coming out of boiler furnace is made to fall over the belt conveyor through a water seal as shown in figure. The cooled ash falls on the belt conveyor and it is carried continuously to the dumping site or overhead bunker. The ash is carried to the dumping site from the ash bunker with the help of trucks.

The control valve is opened and closed manually to load the truck. This type of ash handling system is used at Khapaskheda power station which is near Nagpur.The life of this system is 5 to 10 years. The maximum capacity of this system is limited to 5 tons per hour. The major advantage of this system is low power consumption.

2. Hydraulic ash handling system. The hydraulic ash handling system carries the ash with the flow of water with high velocity through a channel and finally dumped to the sump. The hydraulic system is subdivided as low velocity system and high velocity system.

Low velocity system. The arrangement of the system is shown in Figure 1.51 (*a*). In this system, ash from the furnace grate falls into the system of water possessing low velocity and carried to sump with water. The velocity of water in the water-trough is usually between 3 to 5 m/sec. With the use of higher water velocity, the abrasion is reduced as the ash tends to ride on the water instead of scouring along the bottom. The ash is separated front water when it reaches to the sump. The separated water is used again while the ash collected in the sump is sent out through carriages. The ash carrying capacity of this system is 50 tons/hr and distance covered is 500 meters.

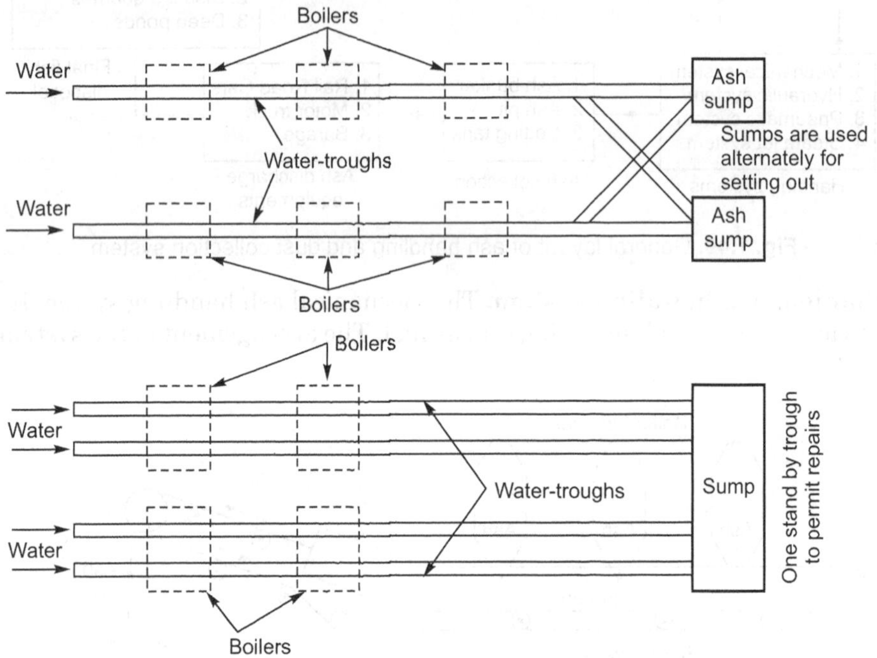

Fig. 1.51. (*a*) Low-pressure Ash Handling Hydraulic System

High velocity system. The arrangement of the system is shown in Figure 1.51 (*b*). The hoppers below the boilers are fitted with water nozzles at the top and on the sides as shown in figure. The top nozzles quench the ash and the side nozzles provide the driving force to carry the ash through a trough. The cooled ash with high velocity water is carried to the sump through the trough as shown in figure. The water is again separated from ash and recirculated. The ash carrying capacity of this system is as large as 120 tons per hour and the distance covered is as large as 1000 meters.

Sometimes the ash and water is pumped to the sump with the help of the pump through the pipe to increase the ash carrying capacity of the system. The only disadvantage is, the pump must be made of special wear resisting material.

The troughs and sumps used in the conveying ash are made of corrosion and wear resistant material as ash carried with water contains many corrosive acids and salts in dissolved form.

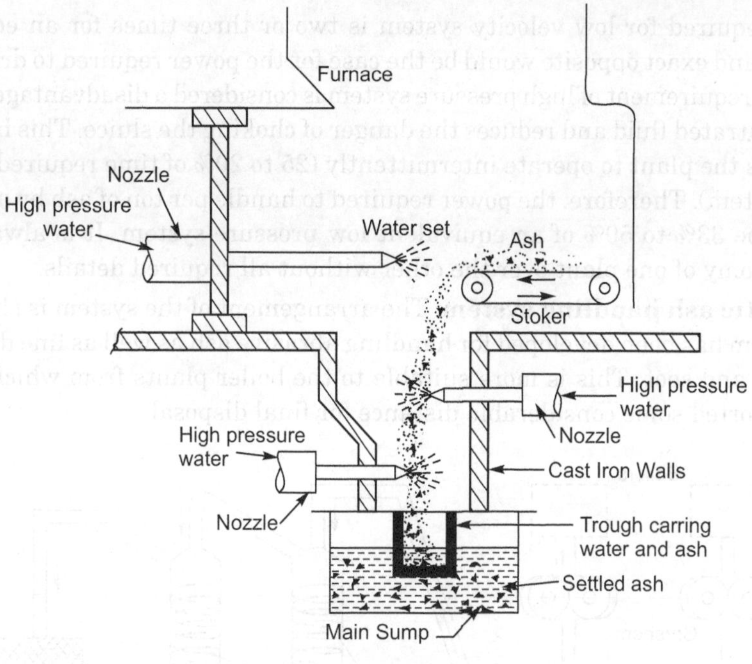

Fig. 1.51. (*b*) High pressure Ash Handling Hydraulic System

The molten slag produced in the pulverised fuel systems can also be carried with the help of high pressure hydraulic system. The cooling nozzles, located near the top of the hopper, direct high pressure water jets against continually running molten stream and disintegrate the slag into small particles as it strikes the surface of water. The formed slag particles fall into a sluice and second set of nozzles transports the slag particles to downcomers discharging into main sluiceway. The slag with water is again discharged to the sump as usual.

In some cases, the water jet pump-type (high pressure high velocity hydraulic system) system is used to pump the ashes or slag to an elevated storage bin installed over a railroad or roadway. The bin is so constructed that the water is drained out and ash or slag is discharged from the bin by gravity flow to rail cars or trucks.

Advantages of Hydraulic System

1. It is clean, dustless arid totally enclosed.
2. Its ash carrying capacity is considerably large therefore it is more suitable (or large thermal power plants.
3. It can discharge the ash at a considerable distance (1000 m) from the power plant.
4. It can also be used to handle a stream of molten ash.
5. The unhealthy aspects of ordinary ash basement work is eliminated.
6. The whole system is clean and healthy.
7. The important feature of the systems is the absence of working parts in contact with the ash.

Low velocity hydraulic system is gaining more popularity in modern thermal power stations as it is simple, able to handle large ash quantities, clean and comparatively economical.

It is always difficult to compare low and high velocity systems without detailed particulars of the two types of plants with water quantities and power requirements.

The water required for low velocity system is two or three times for an equivalent high pressure plant and exact opposite would be the case for the power required to drive the pumps. The high power requirement of high pressure system is considered a disadvantage but it handles high ash concentrated fluid and reduces the danger of choking the sluice. This high rate of ash handling allows the plant to operate intermittently (25 to 20% of time required to continuous low velocity system). Therefore, the power required to handle per ton of ash by a high pressure system would be 33% to 50% of an equivalent low pressure system. It is always difficult to decide the economy of one plant over the other without all required details.

3. Pneumatic ash handling system. The arrangement of the system is shown in Figure 1.52. This system has been developed for handling abrasive ash as well as fine dusty materials such as fly-ash and soot. This is more suitable to the boiler plants from which ash and soot must be transported some considerable distance for final disposal.

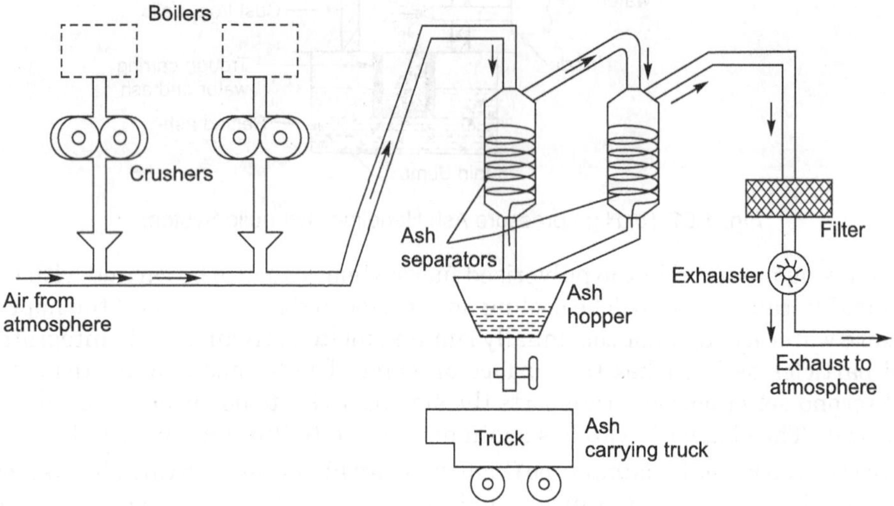

Fig. 1.52. Pneumatic or Vacuum Extraction Ash Handling System

The ash and dust from all discharge points are picked up by a high velocity air stream created by an exhauster at the discharge end. The ash collected in the ash hopper is passed through the ash crushers into the air stream as shown in figure. The ash carried by the air is separated into the primary and secondary separators working on cyclone principle and is collected in ash hopper as shown in figure. The clean air is discharged from the top of the secondary air-separator into the atmosphere through exhauster. The exhauster used may be mechanical as ID. fan or stream jet type or water jet type. If the mechanical exhauster is used, then it is necessary to use filter or air-washer before the air enters into the exhauster to ensure the clean air exhaust to atmosphere. The mechanical exhauster is preferred where large tonnages of material are to be conveyed The power requirement of mechanical exhauster is approximately 3 kW per ton of material. The steam jet exhauster is commonly used for small and medium-sized plant. The steam consumption of this system is approximately 120 kg per ton of material discharged. The water-Jet exhauster may be used more economically where large quantities of water are easily and cheaply available. The ash carrying capacity of this system varies from 5 to 30 tons per hour.

The advantages and disadvantages of this System are listed below:

Advantages

1. With the use of this system, all dust nuisance is eliminated in the handling of flyash and dust. The dustless operation is possible as the materials are handled totally in an enclosed conduit.

2. There is no spillage and rehandling.

3. The materials are conveyed in a dry state and delivered to the storage bin in the same condition. Therefore, there is no chance of ash freezing or sticking in the storage bin and material can be discharged freely by gravity.

4. The system has great flexibility and thus can be made to fit varying physical plant conditions.

5. The conveyor pipeline requires little space in the boiler plant and therefore the cost of the plant pet ton of ash discharged is less than the other systems.

The system suffers only from the disadvantage of wearing out the pipe line and therefore the maintenance charges at a high. It is also more noisy than other systems.

4. Steam jet system. In this system of ash handling, the steam is passed through a pipe at sufficiently high velocity which is capable of carrying dry solid materials of considerable size along with it. The high velocity is given to the steam by forcing the steam through the pipe under pressure greater than that of atmospheric.

In a high pressure steam jet system, a jet of high pressure steam is passed in the direction of ash travel through a conveying pipe in which the ash from the boiler ash hoppers is fed. The ash is deposited in the ash hopper. The advantages and disadvantages of this system are listed below:

Advantages

1. The Steam generated by the boiler is used, therefore, it does not require any sort of Auxiliary drive.

2. Ash can be removed economically by this system through a horizontal distance of 200 meters and through a vertical distance of 30 meters.

3. The capital cost of this system per ton of ash handled is less than other systems.

4. It requires less space.

5. The equipment can be installed in awkward position too.

Disadvanges

1. There is greater wear of pipes carrying the ash due to abrasive action of ash. Therefore, the pipe is lined with nickel alloy.

2. The operation of the system is noisy.

3. The capacity of this system is limited to 15 tons per hour therefore continuous operation of the system is necessary.

1.11 FORCED DRAUGHT

In, a forced draught system, a blower is installed near the base of the boiler and air is forced to pass through the furnace, flues, economizer, air-preheater and to the stack. This draught system is known as positive draught system, or forced draught system because the pressure of air throughout the system is above atmospheric pressure and air is forced to flow through the system. The arrangement of the system is shown in Figure 1.53.

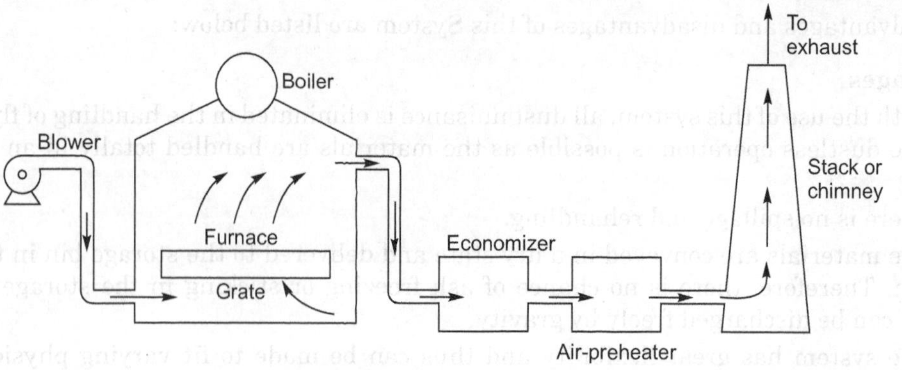

Fig. 1.53. Forced Draught

A stack or chimney is also used in this system as shown in figure but its function is to discharge gases high in the atmosphere to prevent the contamination. It is not much significant for producing draught therefore height of the chimney may not be very much.

Induced Draught

In this system, the blower is located near the base of the chimney instead of near the grate. The air is sucked in the system by reducing the pressure through the system below atmosphere. The induced draught fan sucks the burned gases from the furnace and the pressure inside the furnace is reduced below atmosphere and induces the atmospheric air to flow through the furnace. The action of the induced draught is similar to the action of the chimney. The draught produced is independent of the temperature of the hot gases therefore the gases may be discharged as cold as possible after recovering as much heat as possible in air-preheater and economizer.

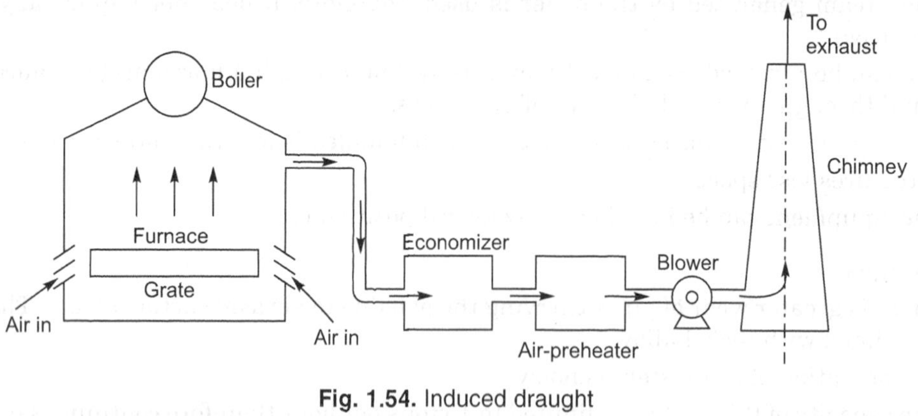

Fig. 1.54. Induced draught

This draught is used generally when economizer and air-preheater are incorporated in the system. The fan should be located at such a place that the temperature of the gas handled by the fan is lowest. The chimney is also used in this system and its function is similar as mentioned in forced draught but total draught produced in induced draught system is the sum of the draughts produced by the fan and chimney. The arrangement of the system is shown in Figure 1.54.

Comparison of Forced and Induced Draughts

The advantages of the forced draught over the induced draught are listed below:

1. The size and power required by the induced draught fan is more than the forced draught because the induced draught fan handles more gases (air and fuel) and at elevated temperature. The volume of the gas handled by induced draught fan is much larger than the volume handled by forced draught fan due to high temperature of the gases, therefore the size of induced draught fan is 1.3 times the size of forced draught fan.

2. Water cooled bearings are required for induced draught fan to withstand the high temperatures of the flue gases.

3. There is no chance of air leakage in the furnace with forced draught as the pressure inside the furnace is above atmospheric pressure. There is continuous leakage of air in the furnace with induced draught as the pressure inside the furnace is less than the atmospheric pressure. This dilutes the combustion.

4. The flow of air through the grate and furnace is more uniform and it penetrates better into the fire bed when forced draught is used. The better penetration of air through the fuel bed and uniform flow improves the rate of burning.

5. When the doors are opened for firing in case of induced draught fan, there will be rush of cold air into the furnace and this reduces the draught through the system and reduces the heat transmission efficiency of the surface.

1.12 SUPERHEATERS

The function of the superheater in the thermal power plant is to remove the last traces of moisture (1 to 2%) from the saturated steam coming out of boiler and to increase its temperature sufficiently above saturation temperature. The superheating raises overall cycle efficiency as well as avoids too much condensation in the last stages of the turbine (below 12%) which avoids the blade erosion.The heat of combustion gases from furnace is utilised for the removal of moisture from steam and to superheat the steam. Superheaters usually have several tube circuits in parallel with one or more return bends, connected between headers.

Heat from the hot gases to the vapour in the superheater is transferred at high temperatures. Therefore, primary section of superheater is arranged in counterflow and secondary section in parallel flow to reduce the temperature stressing of the tube wall. The metal used for superheat tubes must have high temperature strength, high creep strength and high resistance to oxidation as superheater tubes get rougher service than water walls of the modern boilers. Carbon steels (510°C) and chromium-molybdenum alloys (650°C) are commonly used for superheater tubes.

Considerable ingenuity is necessary to provide superheaters for modern steam power plants. The problem of superheater design and location is complicated by the requirements of increased gas temperatures to provide higher steam temperature. In modern boilers, the temperature of combustion is approaching the fusing temperature of ash in the coal and therefore there is a tendency of the to collect in fluid form on the superheater tubes (slagging).

The problem of slagging the superheater is partly eliminated by following methods:

1. Locate the superheater close to the furnace in order to develop the required steam temperature.

2. A bank of screen tubes is located in front of the superheater to limit slag accumulation.

3. Limiting the constant temperature range to 60 to 65% of load rating of the steam generator.

4. With the use of combined convection-radiation superheater.

The steam is superheated by transferring the heat of gases either by convection or by radiation or by combined convection and radiation. Figure 1.55 shows how steam temperature changes with load for each of these types. The combined type superheater has proved most desirable as it keeps the constant temperature throughout the load range.

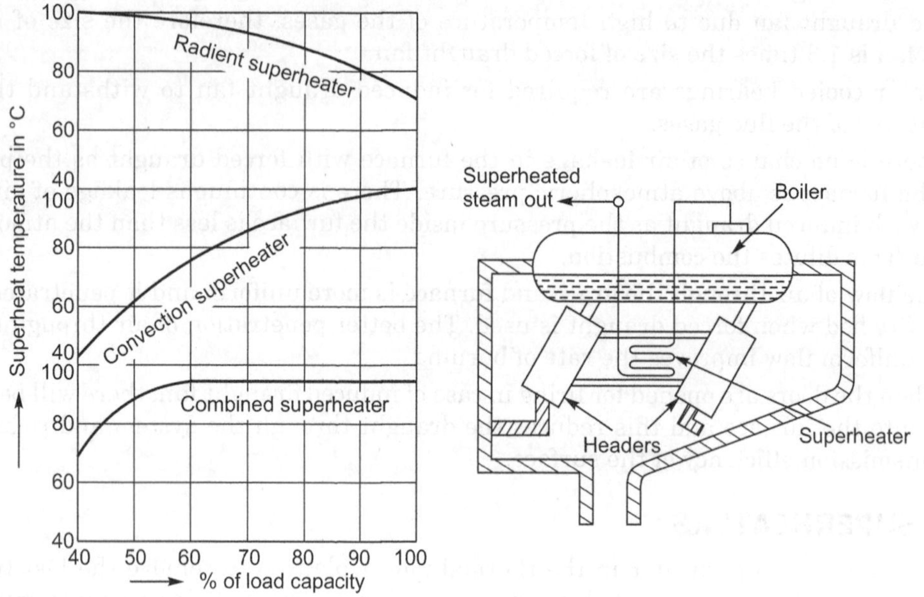

Fig. 1.55. Superheater and steam temperature changes with load

The principle of convection superheater is similar to steam generating tubes of the boiler. The hot gases at high temperature sweep over superheater tubes and raise the temperature of steam which magnitude depends upon exit gas temperature leaving the superheater and gas-velocity.

Dry-wet Combined Cooling Towers

Several different types of closed cycle cooling systems have been proposed to meet the increasing needs of the power plants within the constraints imposed by water resources limitations and environmental restrictions. Whatever be the situation, the need for enlarged make up water supplies for electric power generation is a growing problem to the power industry and, therefore, the answer must be to design cooling systems that make the most efficient use of the available cooling water. One of the most promising of these is the dry-wet combination cooling system.

The combination of dry and wet cooling water (Figure 1.56) is a new approach to reduce the water requirement in the cooling circuit and simultaneously-reduce the working cost of the cooling system (₹/kWh). A 1000 MW nuclear power plant operating at rated load using all-wet system requires about 20 million gallons of make-up water each day (0.83 gallons/kWh). Therefore, the capital cost of water saving with wet-dry system compared with all-wet or all-dry system is frequently an important consideration.

A dry cooling system (only air-cooled) is divided as 'Indirect' and 'Direct' cooling system as mentioned in previous article. In 'direct' system, the exhaust steam from the turbine is condensed directly in the air-cooled condenser. In 'indirect' system, the exhaust steam from the turbine is first condensed with the use of cooling water and heated cooling water is cooled with the help of air (without coming in contact as in heat exchangers) and then it is recirculated back to condenser. Direct and indirect air-cooled cooling systems do not work efficiently during hot summer and are incapable to the fluctuating loads. Conventional wet type system does not work efficiently during cold weather. Power plant cooling capacity employing evaporative type wet towers can be supplemented with dry towers (air-cooling). A new condenser water cooling system is proposed to use both conventional wet towers and conventional dry-towers to meet every conceivable make-up water situation from 100% wet to 100% dry. The wet-tower is designed to do its greatest work during the summer while the dry tower does its greatest work during cold weather. The capability of these two cooling systems is combined to condense the turbine exhaust steam in duel-service condenser. The net result is a new cooling type system that can use conventional low exhaust pressure turbines and make reduced make-up water situation that a utility faces. It will use every available gallon of water for evaporative cooling at the higher ambient temperature conditions and discharge the remaining waste heat directly into the atmosphere as sensible heat. In this way, dry-wet cooling system combines low water use features of dry systems and high cooling efficiency feature of wet system in any operationally flexible combination. Any type of make-up water supply to the wet tower can be used as potable, blackish or sea water.

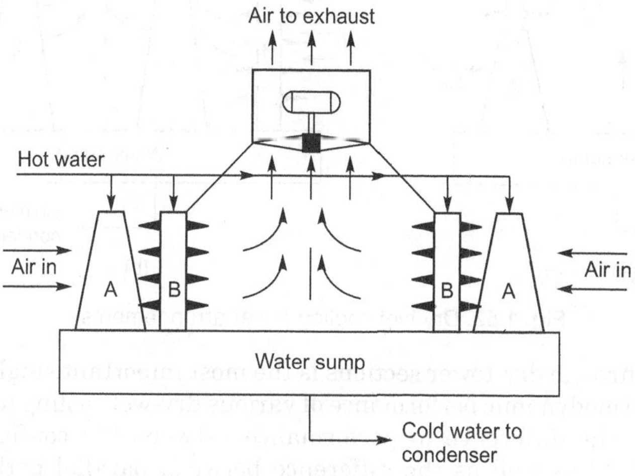

Fig. 1.56. Dry wet combined cooling Tower

There are many different configurations into which dry-tower and wet-tower can be arranged to form a combined dry-wet cooling tower system. Dry-wet cooling tower systems are mainly classified as parallel path and serious path towers according to their air side design.

Arrangement of Dry-Wet Towers: In parallel path towers, the air flows in parallel streams through—the dry and wet sections. This results in a system with a relatively low air pressure drop. Both sections are benefited thermally since each receives cool ambient air.

Combination of dry-wet towers will optimize with larger dry sections in colder climates while higher wet sections will be required in warmer climates. The dry-only and dry-wet modes of operation are more often attractive in colder climates than in warmer and consequently the water evaporation and fogging magnitudes are lower. Furthermore, for a dry-wet tower of a given size, the energy and fuel consumption are reduced in colder climates as compared with warmer climates.

The water evaporation in wet tower and fogging potential are closely related. It is possible to constrain the operation of the dry-wet tower to maintain minimum water loss and prescribed level of fogging.

Increasing the dry surface area achieves more economical cooling since the fogging magnitude limit is not violated. Thus, the total cost of the system drops as the dry surface area is made larger until increasing capital costs offset the decreasing operating costs. Generally, imposition of severe fogging constraints make combination tower with larger dry-sections and higher capital costs economically superior to wet towers.

The selection of basic wet section size depends upon the prevailing meteorological conditions, the optimum size of the dry section to be added to it depends primarily upon the water cost. The increased capital cost of the combination towers is then compensated for by the lower operating costs resulting from water and fuel conservations.

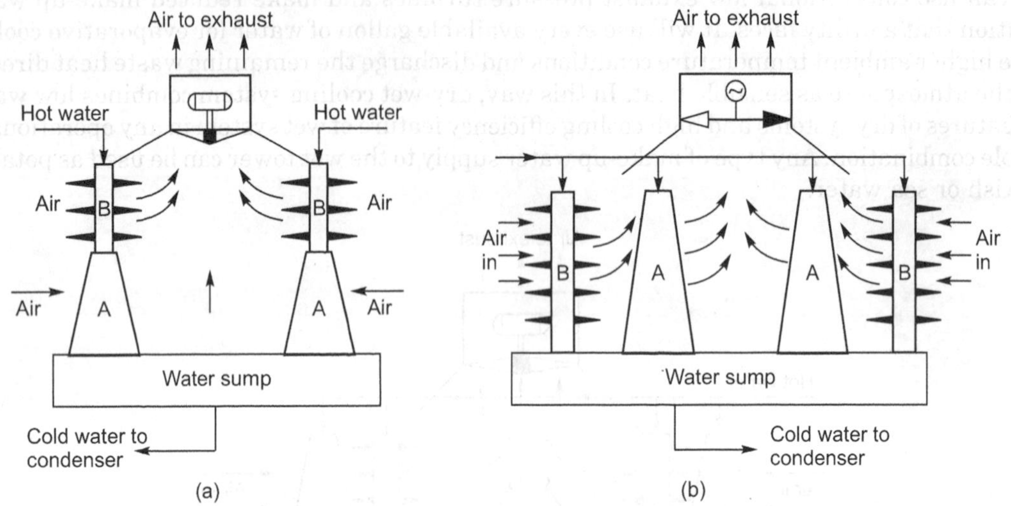

Fig. 1.57. Dry-wet cooling tower arrangements

The air flow rate through dry-tower sections is the most important single factor influencing the economic and thermodynamic performance of various dry-wet cooling tower arrangements. This factor explains the difference in performance between the configurations shown in Figures 1.57 (*a*) and (*b*) as well as the difference between parallel path and serious path configurations. In general, the configuration shown in Figure 1.57 (*b*) is economically superior to that with combined sections except when water conservation or fog abatement is of overriding importance. Parallel path towers are generally far more economical than series path towers.

The dry-wet parallel flow towers currently manufactured appear to offer significant water saving but at a higher capital cost than mechanical draft wet towers with similar capacity. If the dry sections are not sized larger enough, the water saving and fog abatement are achieved at the expense of higher cold water temperatures coming out of tower and therefore poor power plant efficiencies.

Other two arrangements of series and parallel are discussed below. A series arrangement of dry-wet cooling system is shown in Figure 1.58. The water coming out of condenser is first passed through dry-cooling tower, where it is cooled by passing through a finned heat exchanger with the help of air. Then the water coming out of dry tower is passed through wet cooling tower as shown in figure where it is cooled in the direct contact of air by evaporative cooling. The water collected in the basin is supplied to the condenser with the help of pump.

Another parallel arrangement of dry-wet system is shown in Figure 1.58. The load of condenser (condensing of steam) is partly taken by the dry tower and partly by wet tower.

In the right-hand circuit, the water used for condensing steam is cooled in an indirect air-cooler (dry-system) and cooled water is again circulated through condenser with the help of pump. In the left-hand circuit, the water used for condensing the steam is cooled in a direct current of air (wet-system) and cooled wafer collected in the basin is again circulated through the condenser with the help of pump.

In both systems, the water evaporated by the wet-section of the system is compensated from the external source.

The economy of the combined system lies in the capital and working cost. In addition to this, the combined system conserves the water required for the power plant. This system allows the flexibility for locating the power plant at any required place where there is a shortage of water.

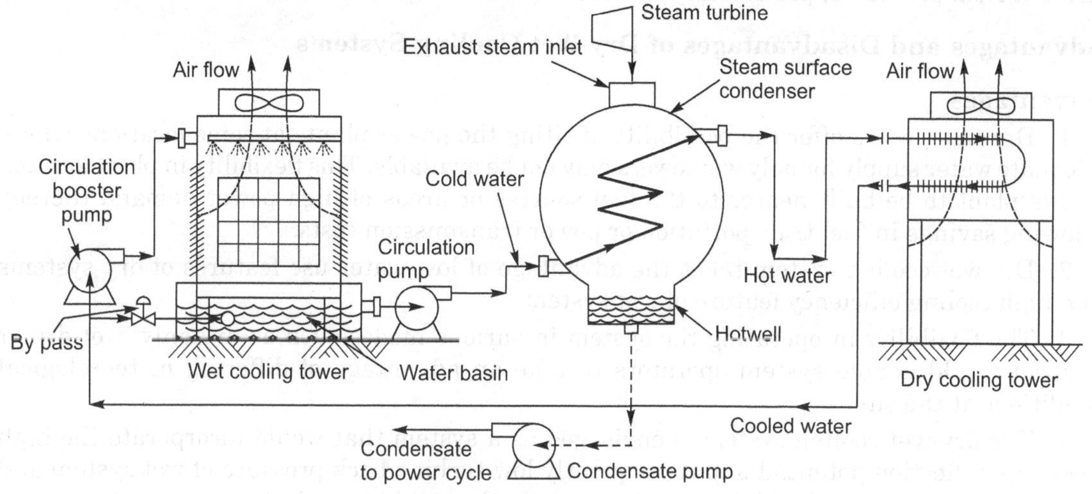

(*a*) Series arrangement of a dry wet system

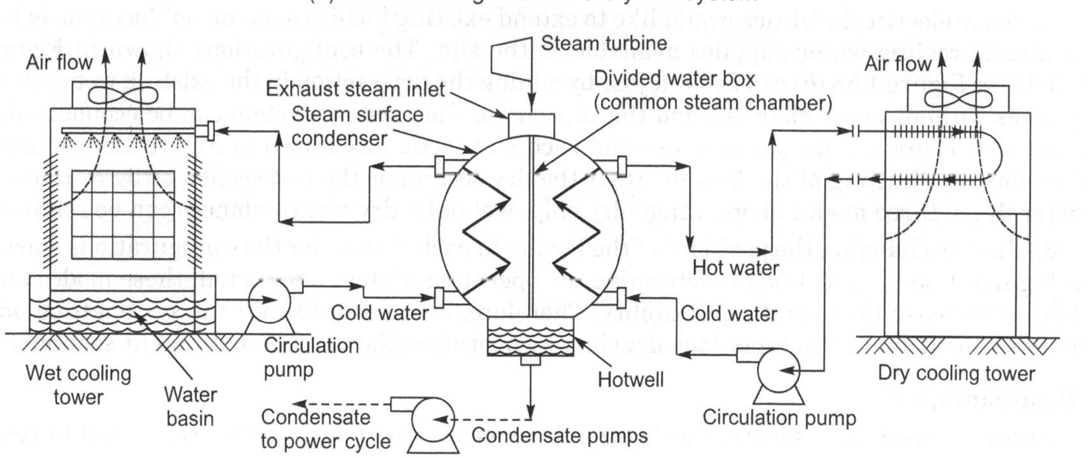

(*b*) Parallel arrangement of dry-wet system

Fig. 1.58

Location of cooling tower. The following points should be carefully considered when locating the cooling tower.

(1) The tower site should be such that it allows unrestricted air flow to the tower. As much open space as far as possible should be allowed between the cooling tower louvers and nearby structures.

(2) It is always advisable to place a cooling tower away from office building, laboratory and control room in the plant because of the possibility of the moist air and noise of the tower causing inconvenience to the working people.

(3) Prevailing wind direction should be studied so as to minimize their circulation effect of hot air from the tower which should be carried away from the tower and should not re-enter through the louvers.

(4) The tower should be so located that the piping runs to and from it should be minimum. The tower should be as near as possible to the source of make-up water.

(5) The placement of the tower should also be studied with reference to nearby chimney and effluent from processes, particularly if these are hot or oily.

Advantages and Disadvantages of Dry-Wet Cooling Systems

Advantages

1. Dry-wet towers offer the possibility of siting the power plants at some locations where adequate water supply for only wet towers may not be available. This flexibility in plant location allows plant to be built nearer to the fuel sources or areas of high power demand thereby allowing savings in fuel transportation or power transmission costs.

2. Dry-wet cooling system takes the advantage of low water use features of dry systems and high cooling efficiency feature of wet system.

3. The flexibility in operating the system in various modes such as dry only, wet-dry or dry-wet would enable system operators to take an advantage of differing meteorological conditions at the site.

4. The dry-wet cooling system is envisaged as a system that would incorporate the high heat rate reflection potential and consequently low turbine back pressure of wet system and yet would not result in the high water losses and other problems of wet system as fogging.

5. Some electrical utilities would like to extend existing facilities but do not have the extra additional cooling water supplies available at the site. The configurations shown in Figure 1.58 (*a*) or Figure 1.58 (*b*) can be built just by adding the dry-section in the existing wet cooling systems without many changes, and the capacity of the existing system can be economically increased. There is a fan for each dry and each wet section as shown in Figure 1.58 (*a*) and, therefore, by shutting of the fans in either the dry section or the wet section or by re-routing water flows, three modes of operation (dry only, wet only, dry-wet combined) can be realized.

6. Just by changing the positions of the towers in each section for the configurations shown in Figures 1.58 (*a*) and 1.58 (*b*) determine the operation of the system in all these modes and thereby increase the operating flexibility. Therefore, it is easier to shift the system from one combination (say wet) to other (say dry) just by operating the louvers in different sections.

Disadvantages

Dry-wet towers have some of the drawbacks associated with both dry as well as wet towers.

1. The capital costs of dry-wet towers are higher than similar duty wet towers.

2. There are start-up and shut-down problems as freezing is commonly associated with dry towers and therefore special arrangements are required to avoid this.

3. There are still some water consumption, fogging and related problems similar to wet towers.

4. To evaluate the proposed combination of dry-wet cooling system with regard to the environmental consequences, it is necessary to calculate water consumption and fogging magnitude at all meteorological conditions that can occur at proposed site. Both capital and

operating costs are usually estimated by assuming a set of 'fixed' design conditions, expected to prevail for small fraction of the working period. Such fixed *conditions* designs would not realistically incorporate the advantages of inherent flexibility of dry-wet tower and would not adequately evaluate operating expenses that prevail at all of the *off-design* meteorological conditions.

Steady increasing fuel costs, constrains on permissible environmental inputs and limitations on water availability in addition to the national objective of efficient use of fuel and water make it increasingly necessary to optimize the whole heat rejection system of the power plant and to make dry-wet cooling system more economical than wet system alone.

1.13 REGENERATOR

In the closed cycle system, the heat of the turbine exhaust gases will be again utilized. The hot gases will not be allowed to go as waste. To make use of this heat, a regenerator is provided in the system (Figure 1.59). In the regenerator, the heat of the hot exhaust gases from the turbine is used to preheat the air entering the combustion chamber.

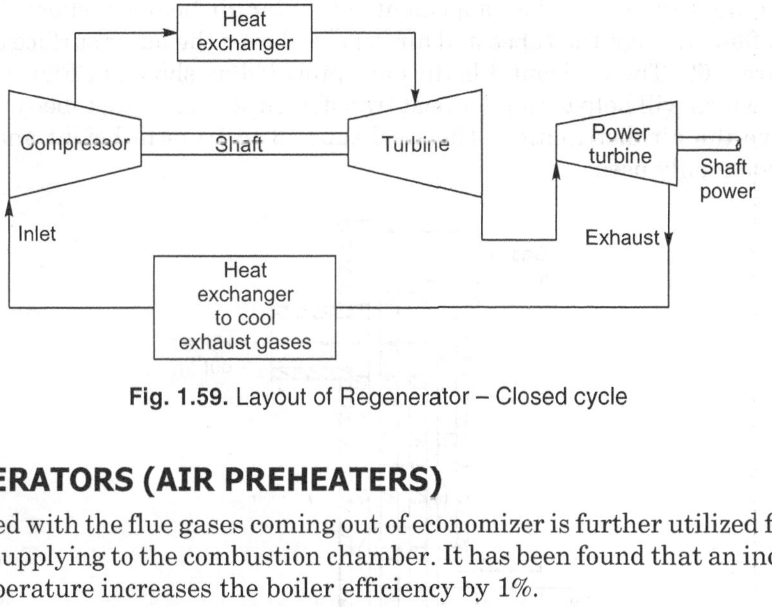

Fig. 1.59. Layout of Regenerator – Closed cycle

1.14 DEAERATORS (AIR PREHEATERS)

The heat carried with the flue gases coming out of economizer is further utilized for preheating the air before supplying to the combustion chamber. It has been found that an increase of 20°C in the air temperature increases the boiler efficiency by 1%.

The air heater is not only considered in terms of boiler efficiency in modern power plants, but also as a necessary equipment for supply of hot air for drying the coal in pulverized fuel systems to facilitate grinding and satisfactory combustion of fuel in the furnace. The use of preheater is much economical when used with pulverized fuel boilers because the temperature of flue gases going out is sufficiently large and high air temperature (250 to 350°C) is always desirable for better combustion.

Air heaters are usually installed on steam generators that burn solid fuels but rarely on gas or oil fired units. By contrast, economizers are specified for most boilers burning liquid or gas or coal whether or not an air heater is provided. The principal benefits of preheating the air are (1) improved combustion, (2) successful use of low grade fuel (high ash content) (3) increased thermal efficiency, (4) saving in fuel consumption and (5) increased steam generation capacity (kg/m^2-hr) of the boiler.

The air-preheater must provide reliability of operation, should occupy small space, must be reasonable in first cost and should be easily accessible. The air-heaters are not essential for the operation of steam-generator but they are used where a study of the costs indicates that some money can be saved or efficient combustion can be obtained by their use. The decision for its adoption can be made when the financial advantage is weighed against the capital and maintenance cost of the heater. The decision cannot be taken so easily as the economic advantages of hot combustion extend to the size of the boiler, the efficiency of combustion, maintenance of furnaces and the saving in heat discharged to the chimney.

The different types of air-heaters which are in use are discussed below:

The air-preheaters are generally divided into two groups as recuperative and regenerative type. The recuperative heaters continuously transfer the heat from hot gases to cold air. The regenerative heater alternately gets heated and cooled by the hot gases (absorbing heat) and cold air (dissipating heat). Unlike the recuperative type, the regenerative is discontinuous in action and operates on cycle. In rotary regenerative type, the cyclic action applies to the heating and cooling of an individual element of the surface but the flowing stream of air receives heat continuously.

The two recuperative types of heat-exchangers which are commonly used for air-heating are described below:

Tubular air-heater. A typical arrangement of tubular air-heater is shown in Figure 1.60. The flue gases flow through the tubes and air is passed over the outer surface of the tubes as shown in Figure 1.60. The horizontal baffles are provided as shown in Figure 1.60 increase time of contact which will help for higher heat transfer. In some design, tube-row staggering is used to improve the air-distribution. The steel tubes 3 to 10 m in height and 6 to 8 cm in diameter are commonly used.

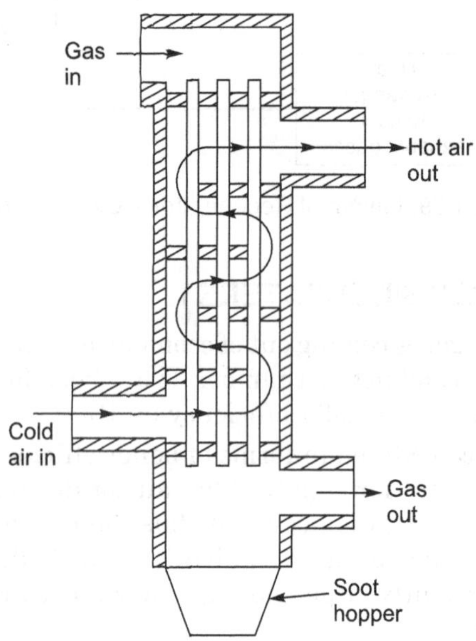

Fig. 1.60. Tubular air-heater

Plate Type Air-heater. A plate type heater is shown in Figure 1.61 (*a*). It consists of rectangular flat plates spaced from 1.5 to 2.5 cm apart leaving alternate air and gas passages. This type of air-heater is not used in modern installations as it is more expensive both as to flat cost and maintenance cost compared with tubular air-heaters.

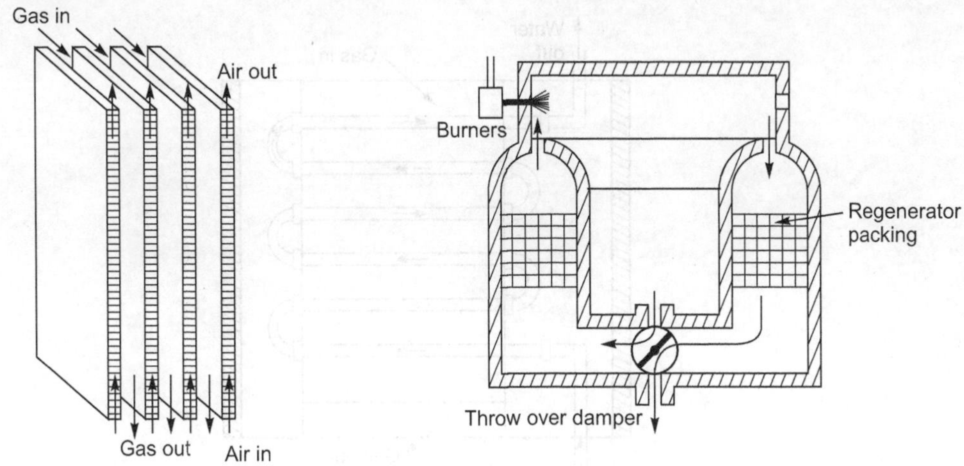

(a) Plate type air preheater (b) Regenerative heat exchanger.

Fig. 1.61

Regenerative heat exchangers. A typical type of stationary regenerative heat exchanger is shown in Figure 1.61 (*b*). The transfer of heat from hot gases to cold air is divided into two stages. In the first stage, the heat of the hot gases flowing through the heat-exchanger is transferred to the packing of the heater and it is accumulated in the packing and the hot gases are cooled to sufficiently low temperature before exhaust to atmosphere. This stage is referred to as "Heating period". In the second stage, the cold air is passed through the hot packing where the heat is accumulated and the heat from the packing is transferred to the cold air. This stage is known as "Cooling period".

Two such are required in stationary regenerative heater as shown in figure. The period of heating, the period of cooling, the amount of packing required depends upon the mass of gas and air flowing through the packing, the initial and final temperatures of the gases and rise in temperature of the air required.

A rotary type regenerative heat exchanger is described in the chapter of gas turbine power plants so the readers are requested to see the same. For continuous operation of airheater, an arrangement must be made to clean the heating surface, particularly gas side, This is done in practice by providing openings for the use of brushes and steam lances or by permanent soot blowers. Hoppers are also provided to accumulate soot at the bottom of the heater.

1.15 ECONOMIZER

The economizer is a feed water heater deriving heat from the flue gases discharged from the boiler. The justifiable cost for economizer depends on the total gain in efficiency. In turn this depends on the gas temperature out of the boiler and feed water temperature to the boiler. Regenerative cycle inherently, gives high feed water temperature. Therefore the adoption of economizer must be studied very carefully. A typical return bend type economizer is shown in Figure 1.62.

A boiler producing between 10 to 100 tons of steam per hour and operating at 30% or more load should be evaluated for possible retrofitting with an economizer. The cost benefits depend upon the boiler size, type of fuel used and exhaust gas temperature. It has been estimated that about 1% of fuel costs can be saved for every 6°C rise in the temperature of the boiler feed water. Saving up to maximum 20% can be achieved by incorporating economizer where boiler operates very efficiently.

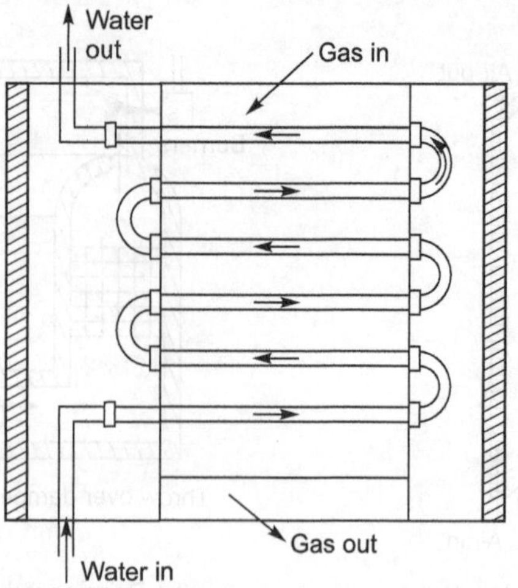

Fig. 1.62. Typical return bend type economizer

When more heat is available, that can be used in increasing the sensible heat of the feed water or pass it through an air heater. However, in most economizers, the feed water is not heated higher than to within 25°C of the temperature corresponding to the saturation temperature of steam in the boiler thus preventing steam formations in the economizer.

1.16 COOLING TOWERS

1. Natural draft spray filled tower. The arrangement of this type of tower is shown in Figure 1.63. The air enters through the lovered sides and flows across the unit in a transverse direction. The air circulation through the tower depends on wind velocity. The capacity of this tower varies from 50 to 100 litres per minute per m^2 of base area depending upon the air velocity.

These towers are used only for diesel plants and where prevailing winds are not cut-off by obstructions. This is not used for high capacity thermal plants as cooling range is limited, wind losses are high and there is no control over the outlet temperature of water.

2. Natural draft packed type tower. The construction of this tower is similar to spray filled tower except that water distributing, toughs of fills are used which helps to break the water into small droplets. In this tower (Figure 1.64) also, the flow of air is cross-wise to the flow of water.

These towers are also rarely used for thermal power plants as original cost (due to height requirement) and pumping head required are high. Tower's extreme length and height and narrow width require anchoring to withstand high winds.

3. Hyperbolic cooling tower. The first hyperbolic natural draft reinforced concrete tower was designed by Prof. Van Itesson of Dutch state and installed at Emma Collisey in 1916:

The hyperbolic cooling towers are widely used in Europe. First unit of this type was installed in U.S.A. at Big-Sandy station of Kentucky Power Co. It is capable of handling 120 × 10^3 gpm and cools the water from 43°C to 30°C. It has minimum diameter of 39.5 m and maximum diameter of 74.5 m and is 400 meters high. It serves to the station of 265 MW capacity. Another notable cooling tower of this type is Gundremmingen, 160 m high, 135.2 m bottom and 83.2 m top diameter established in 1978 at Germany.

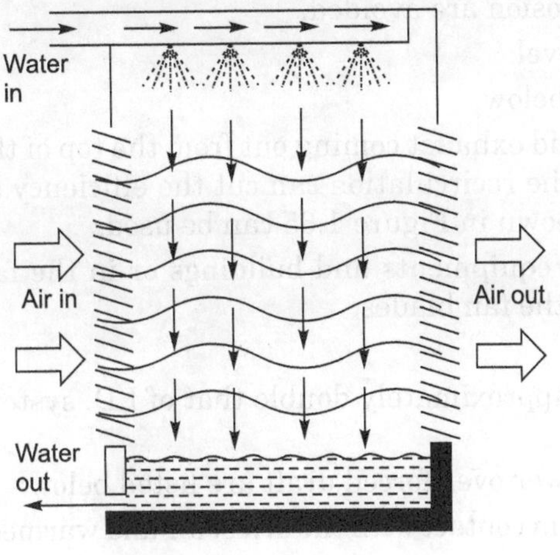

Fig. 1.63. Natural draft spray filled tower

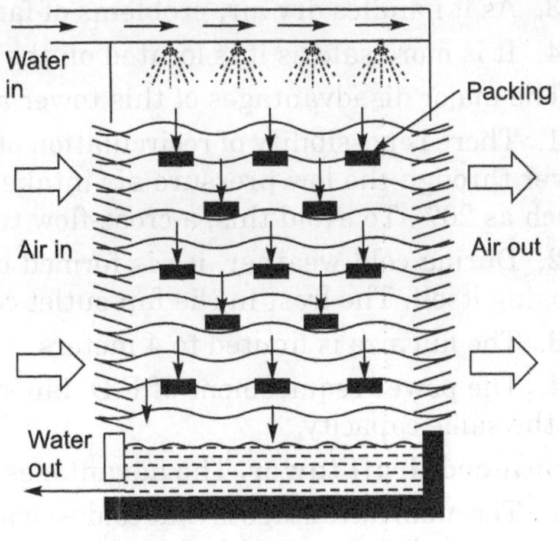

Fig. 1.64. Natural draft packed type tower

It is steel reinforced concrete structure mostly slack (empty space) and the bottom 10 m above the air-intake contains packing over which warm water flows. The shape of the stack is circular in plan and hyperbolic in profile. The operation of this tower is much like that of other natural draft spray cooling towers with hot water cascading over timber splash type filling through which cooler air moves.

Any cooling tower breaks the warm water into a spray or a film of water and expose it to a flow of air. Some of the heat in the water is directly transferred to the air stream and some of the water evaporates, providing significant cooling effect. Then the moist warm air goes to the atmosphere and the cooled water is collected and returned to the power plant.

Mechanical Draft Cooling Towers

The natural draft towers are totally replaced by mechanical draft towers as they provide closer approach to WBT, give higher efficiency, reduce spray and windage losses and require reduced ground area.

Mechanical draft towers require less space and less piping than natural draft towers. Lower water temperatures supplied by mechanical draft towers usually boost overall plant economy enough to cover the added operating charges and higher initial cost of installation.

The mechanical draft towers use fans to move the air through the tower instead of depending on natural draft or wind velocity. This speeds cooling and increases the efficiency of tower by increasing the air-velocity over wet surfaces and through the tower. The correct amount of air at required velocity can be readily selected to meet job cooling requirements. With the use of mechanical tower, much smaller equipment can be used to handle air-flow under fairly well controlled operational conditions.

The mechanical towers are subdivided as forced draft, and induced draft. The induced draft is further divided as counter flow and cross-flow types.

Comparison of Forced and Induced Draft Towers

Forced draft towers. The advantages of this type of towers are listed below:

1. This is more efficient than induced draft as some of the air velocity is converted into static pressure in the tower and recovered in the form of useful work.

2. The vibration and noise are minimum as mechanical equipments are set on a solid foundation.

3. As it handles dry air, problems of fan blade erosion are avoided.
4. It is more safe as it is located on the ground level.

The major disadvantages of this tower are listed below:

1. There is possibility of recirculation of hot, humid exhaust coming out from the top of the tower through the low pressure air intake region. The recirculation can cut the efficiency as much as 20%. To avoid this, a cross-flow tower as shown in Figure 1.65 can be used.
2. During cold weather, ice is formed on nearby equipments and buildings or in the fan housing itself. The frost in the fan outlet can brake the fan blades.
3. The fan size is limited to 4 meters.
4. The power requirement of F.D. fan system is approximately double that of I.D. system for the same capacity.

Induced draft tower. The advantages of this tower over forced draft are listed below:

1. The main advantage is that coldest water comes in contact with the driest air and warmest water comes in contact with the most humid air.
2. The recirculation is seldom a problem with this tower as outlet fan discharges the heated and humid air directly away from the air-intakes below the tower.
3. The size of 20 m in diameter can be used.
4. Claims are made that the I.D. fan tower has the advantages of lower first cost, requires less space, is capable of cooling through a wide range.
5. The first cost is lower due to the reduction in pump capacity required and smaller length of water pipes. The power consumption per kg of water cooled is less compared with F.D. fan system.

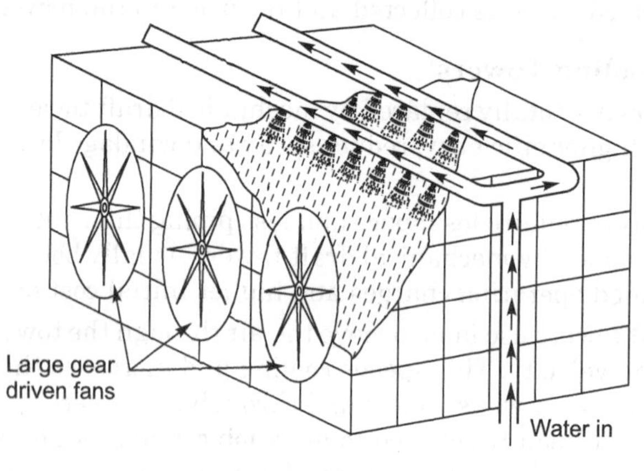

Fig. 1.65. Cross-flow tower

1.17 SELECTION OF STEAM POWER STATION

To get an overall economy we should take care of the following factors while choosing a site for steam power station

***(i)* Supply of fuel:** The plant should be located in a place near coal mines hence the transportation cost could be reduced. If it is not able to construct a plant near the coal mines or the fuel source at least it must be considered that there are proper facilities available for the transportation of fuel ie coal to the generating station.

***(ii)* Availability of water:** A large amount of water is required for the condenser for proper cooling process so there should be natural resource nearby the plant to get continuous supply of water. It could be a river, stream or lake any water source nearby.

***(iii)* Transportation facilities:** A modem steam power station requires the transport of coal as well as the machineries needed for the station. To make this facility available there should be enough roads and rails connecting generating station to different part of the country.

***(iv)* Cost and type of land:** Since the steam power station requires a large area the cost of the land that we buy should be very cheap else it will increase the initial cost to a large amount. And the bearing capacity of the soil should be fair enough to withstand the weight of large machines.

***(v)* Nearness to load centres:** The plant should be located near to the load centers since it could reduce the transmission cost to a very large amount. It is much important if we have to transport D.C. power because it requires high transmission cost, but if we have to transmit A.C. power it could be transmitted a long distance without much cost and so this method could be adopted if the load centers are too far.

***(vi)* Distance from populated area:** The plant should be located at a favorable distance from the populated area because daily tonnes of coal need to be burnt and the smoke and fumes produced *b* it will pollute the atmosphere so the people living nearby area may get lung diseases and other diseases affecting chest cavity. So the plant needs to be placed far away from the living areas.

REVIEW QUESTIONS

Two Marks Questions

1. What is meant by Power Plant?
2. What are the classifications of power plant?
3. What are the Sources of Energy?
4. What are the examples for conventional sources of energy or non-renewable energy?
5. What are the examples for non-conventional sources of energy or renewable energy?
6. What are the factors of power plant performance?
7. What is drift? How is the drift eliminated in the cooling tower?
8. What are the comparisons of sources of energy?
9. State the advantages and disadvantages of pulverized fuel firing.
10. What is thermodynamics?
11. Differentiate heat and Temperature.
12. What are the modes of heat transfer?
13. State laws of thermodynamics.
14. What is the use of load curves in powerplant?
15. State zeroth law of thermodynamics.
16. State first law of thermodynamics.
17. State law of conservation of energy.
18. State entropy and second law of thermodynamics.
19. State third law of thermodynamics.
20. What are the thermodynamic processes?
21. Define supercritical boilers.
22. List the non-flow process.
23. What is thermodynamic cycle?

24. List the most common cycles.
25. Draw the *PV*-diagram of Rankine cycle and mention the process involved.
26. What are the processes in Rankine cycle?
27. What is meant by air standard cycle?
28. What are the assumptions made for air standard cycle analysis?
29. What is Carnot cycle and draw the *PV* and *TS*-diagram?
30. Write short notes on:
(*i*) Reheat cycle (*ii*) Regenerative cycle
31. What is the advantage of reheat cycle?
32. Where the reheat regenerative cycle is used?
33. What is cyclone separator?
34. What are the major components of thermal power plant or draw the line diagram of thermal power plant?
35. Mention the various processes of the Brayton cycle.
36. What is the necessity of feed pump in thermal power plant?
37. List out the major advantages of high pressure boilers in modern thermal power plants.
38. What are the function of the following:
(*i*) Boiler (*ii*) Turbine
(*iii*) Condenser (*iv*) Pump
39. How the boilers are classified?
40. What are the requirements of the boiler?
41. What are the boiler mountings and accessories?
42. Give example for:
(*i*) Low Pressure Boiler (*ii*) High Pressure Boiler
43. Differentiate water tube and fire tube boiler.
44. How steam turbines are classified?
45. Differentiate impulse and reaction turbine.
46. How condensers are classified?
47. Differentiate Jet condenser and Surface condenser.
48. What are the types of pumps?
49. What are the major parts of reciprocating pump?
50. What are the major parts of the centrifugal pump?
51. Sketch a *p-V* and *T*-suitable diagram of Atkinson cycle. Also mention the processes involved in this.
52. What are the types of casings commonly adopted?
53. What is meant by pulverized coal?
54. What are the factors which affect the performance of the pulverized coal burner?
55. What are the types of burner?
56. What is FBC System?
57. What are the types of FBC System?
58. What are the advantages of FBC system?
59. What is meant by coal handling system?
60. What are the types of coal handling system?
61. What is ash handling system?
62. How ash handling systems are classified?
63. Mention various modern ash handling systems.

64. What are the advantages of hydraulic handling system?
65. What are the advantages of pneumatic or vacuum ash handling system?
66. What are the types of draughts?
67. Distinguish between forced and induced draughts.
68. Describe the following:
    (*i*) Super Heater (*ii*) Deaerator
    (*iii*) Heat exchanger (*iv*) Economizer
69. What is cooling tower?
70. What are the steam powerplant cycles?
71. What are the advantages of superheated steam?
72. List down the processes involved in stirling cycle. Mention any four advantages of stirling cycle.
73. Write down the applications of stirling cycle.
74. How cooling towers are classified?
75. What are the factors affecting cooling of water in cooling tower?
76. Why does cooling water flow inside the tubes and steam condense outside of the tubes of the surface condenser?
77. What are the parameters to be considered during the selection of steam power plant?
78. List the thermal power plant in Tamil Nadu.

Sixteen Marks Questions

1. Explain the types of sources of energy and list out the advantages and disadvantages.
2. Explain the models of heat transfer with suitable diagram.
3. Explain the laws of thermodynamics and its limitation.
4. Explain the various types of thermodynamic cycles with *PV* and *TS* diagram.
5. Explain the working principle of steam power plant with suitable diagram.
6. Explain Carnot cycle with *PV* and *TS* diagram.
7. Explain with the schematic sketch the working of different circuit of thermal powerplant.
8. What is Brayton cycle? Explain the working principle of gas turbine power cycle.
9. Explain reheat cycle with suitable diagram and list out the advantages and disadvantages.
10. What is regenerative cycle? Explain with neat diagram.
11. Name the various equipment used for transferring the coal. Explain the working principle of always on equipment.
12. Explain the working principle of steam or thermal power plant with suitable sketch.
13. Draw a neat diagram of cyclone burner and describe its working. What are the outstanding features compared with other burners?
14. Explain with neat sketch the fluidized bed combustion system.
15. Explain the boiler mountings and accessories with suitable sketch.
16. Draw a Rankine cycle for a coal fired and steam thermal power plant. State the means of increasing the efficiency of powerplant.
17. Explain the construction and working principles of following water tube or low pressure boilers:
    (*i*) Babcock Boiler and Wilcox Boiler (*ii*) Cochran Boiler
18. Explain the constructing and working principle of following high pressure boilers:
    (*i*) Benson Boiler (*ii*) Lamont Boiler
    (*iii*) Loeffler Boiler
19. Distinguish between water tube boiler and fire tube boiler.
20. Explain working principle of the impulse and reaction turbine.
21. Draw the pressure velocity diagram of impulse and reaction turbine and explain.

22. Explain the working principles of simple steam condensity plant with sketch.

23. Explain the following with suitable sketch:

(*i*) Jet Condenser (*ii*) Ejector Condenser

(*iii*) Surface Condenser (*iv*) Evaporative Condenser

24. Explain the working principle of the following with neat diagram:

(*i*) Single acting reciprocating pump

(*ii*) Double acting reciprocating pump

(*iii*) Centrifugal Pump

25. Explain the coal burner with suitable sketch.

26. Explain the working principle of FBC system with suitable sketch.

27. Describe the advantages of FBC system.

28. Explain the working principle of coal or fuel handling system with sketch.

29. Explain the ash handling system with suitable sketch.

30. What is meant by draught? Explain the types of draughts.

31. Describe the operation of a balanced draught system with neat sketch.

32. Explain the following with neat sketch:

(*i*) Super heater (*ii*) Air Preheater

(*iii*) Economizer (*iv*) Regenerator

33. What are the types of cooling towers? Explain any one of them with neat sketch.

34. Explain the criteria for selection of steam power station.

2 Hydroelectric Power Plant

2.1 INTRODUCTION

Hydroelectricity is the term referring to electricity generated by hydropower; the production of electrical power through the use of the gravitational force of falling or flowing water. It is the most widely used form of renewable energy, accounting for 16 percentage of global electricity generation.

Most hydroelectric power comes from the potential energy of dammed water driving a water turbine and generator. The power extracted from the water depends on the volume and on the difference in height between the source and the water's outflow. This height difference is called the head. The amount of potential energy in water is proportional to the head. A large pipe (the "penstock") delivers water to the turbine.

"Run of the river" systems do not require a dam or storage facility to be constructed. Instead they divert water from the stream or river, channel it in to a valley and drop it in to a turbine via a pipeline called a penstock. The turbine drives a generator that provides the electricity to the local community. By not requiring an expensive dam for water storage, run-of-the-river systems are a low-cost way to produce power. They also avoid the damaging environmental and social effects that larger hydroelectric schemes cause, including a risk of flooding. Water from the river is channeled through a settling basin, which helps to remove sediment that could harm the turbine. The water then flows into the Fore bay Tank where it is directed downhill through a pipe called a penstock. When the water reaches the bottom, it drives a specially designed turbine to produce the electricity.

Hydroelectric power plants are the most efficient means of producing electric energy. The efficiency of today's **hydroelectric plant** is about 90 percentage.

2.2 CLASSIFICATIONS

Based on the Rating of the Plant

This is the main consideration while assigning a category to a power plant. The rating of the plant will work as the name of the plant especially in case of hydropower plant. Based on rating plants are classified as:

- **Micro hydro power plant:** If the rating of the plant or we can say the output of the plant is less than 100 kW or 0.1 MW, the plant is called Micro Hydro Power Plant.
- **Mini hydro power plant:** If the rating of the plant or we can say the output of the plant is more than 100 kW or 0.1 MW and is less than 1 MW, the plant is called Mini Hydro Power Plant.

- **Small hydro power plant:** If the rating of the plant or we can say the output of the plant is less than 10 MW, the plant will come under the Category of Small Hydro Power Plant. According to the new classification of CEA, India small hydro power plant may have rating up to 25 MW.

Based on the Technology Used

Based on the technology involved specially during the construction of dam and height of the dam and demand the hydro power plants are again classified into three categories. These are:

- **Impoundment hydro power plant:** It is the most common type of hydro power plant based on technological classification. A dam is constructed and water is stored behind dam and the stored potential energy of water behind dam is used as fuel for hydro power plant.
- **Diversion hydro power plant:** In case the water head is small and the potential energy of water is not enough to drive a turbine then a separate water flow is taken from river with high speed so that it may produce enough current in the generator.
- **Pumped hydro power plant:** Sometimes when the demand of power supply is not high then produced electricity is used to drive a motor and water is pumped back to the dam and the potential energy of this water is used to generate the electricity at the plant.

Based on Load Sharing by Hydro Power Plant

- **Peak load power plant:** There types of power plant supply power to the load when there peak load period, only. Rest of the time the power is supplied by main plant. In this types of plants a main power plant is always required and hydro power plant work as secondary plant in this case and share the load for two or three hours.

  Let we have a power plant with real load 120 MW for a period of one hour and rest of the time the load remains within a limit upto 90 MW. In this case installing a main plant of 90 MW with a support hydropower plant of capacity 35 MW is a good idea to save money on construction of main plant.
- **Base load hydro power plant:** This is the power plant which work independently and supply power to the whole load. It work for the whole time *i.e.*, It supply power when there is a requirement. This type of plant are installed where huge water is available.

Based on the availability of head

(*i*) High head power plants (head > 100 m)

(*ii*) Medium head power plants (30 m to 100 m)

(*iii*) Low head power plants (head < 30 m)

Based on the quality of water available

(*i*) Run-of-river plant without pondage.

(*ii*) Run-of-river plant with pondage.

(*iii*) Plant with storage reservoirs.

(*iv*) Pumped storage plants.

(*v*) Mini and micro hydro plants.

2.3 LAYOUT OF HYDROELECTRIC POWER PLANT

Hydroelectric power plants convert the hydraulic potential energy from water into electrical energy. Such plants are suitable where water with suitable *head* are available. The layout

(Figure 2.1) covered in this article is just a simple one and only cover the important parts of hydroelectric plant. The different parts of a hydroelectric power plant are:

1. **Water reservoir:** It stores the water received from the catchment areas during monsoon period. Water surface in the storage reservoir is known as head race.

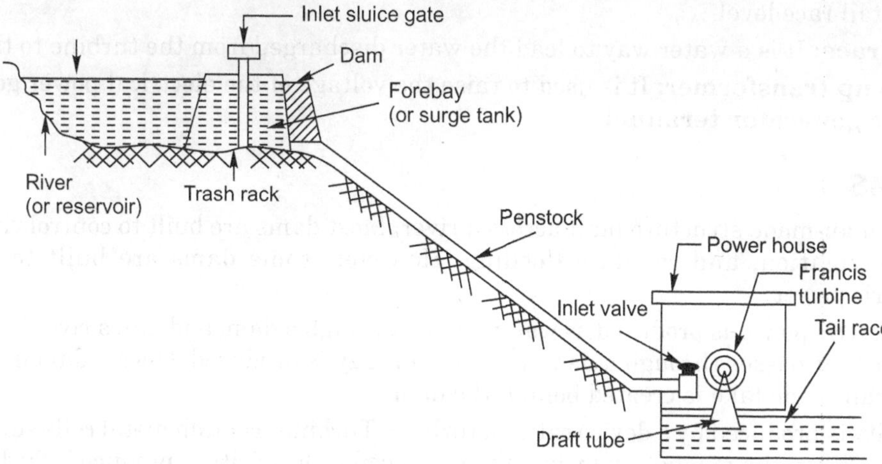

Fig. 2.1. Layout of Hydroelectric Power Plant

2. **Dam:** Dams are structures built over rivers to stop the water flow and form a reservoir. The reservoir stores the water flowing down the river. This water is diverted to turbines in power stations. The dams collect water during the rainy season and stores it, thus allowing for a steady flow through the turbines throughout the year. Dams are also used for controlling floods and irrigation. The dams should be water-tight and should be able to withstand the pressure exerted by the water on it. There are different types of dams such as arch dams, gravity dams and buttress dams. The height of water in the dam is called head race.
3. **Spill way:** A spillway as the name suggests could be called as a way for spilling of water from dams. It is used to provide for the release of flood water from a dam. It is used to prevent over toping of the dams which could result in damage or failure of dams. Spillways could be controlled type or uncontrolled type. The uncontrolled types start releasing water upon water rising above a particular level. But in case of the controlled type, regulation of flow is possible.
4. **Pressure tunnel:** It carries water from the reservoir to surge tank.
5. **Pen stock and tunnel:** Penstocks are pipes which carry water from the reservoir to the turbines inside power station. They are usually made of steel and are equipped with gate systems. Water under high pressure flows through the penstock. A tunnel serves the same purpose as a penstock. It is used when an obstruction is present between the dam and power station such as a mountain.
6. **Surge tank:** Surge tanks are tanks connected to the water conductor system. It serves the purpose of reducing water hammering in pipes which can cause damage to pipes. The sudden surges of water in penstock is taken by the surge tank, and when the water requirements increase, it supplies the collected water thereby regulating water flow and pressure inside the penstock.
7. **Power station:** Power station contains a turbine coupled to a generator. The water brought to the power station rotates the vanes of the turbine producing torque and rotation

of turbine shaft. This rotational torque is transferred to the generator and is converted into electricity. The used water is released through the *tail race*. The difference between head race and tail race is called gross head and by subtracting the frictional losses we get the net head available to the turbine for generation of electricity.

8. **Draft tube:** It is connected to the outlet of the turbine. It allows the turbine to be placed over tail race level.
9. **Tail race:** It is a water way to lead the water discharged from the turbine to the river.
10. **Step up transformer:** It is used to raise the voltage of the electrical power generated at the **generator terminal.**

2.4 DAMS

A dam is a man-made structure built across a river. Most dams are built to control river flow, improve navigation, and regulate flooding. However, some dams are built to produce hydroelectric power.

Hydroelectric power is produced as water passes through a dam, and into a river below. The more water that passes through a dam, the more energy is produced. Once a dam is built, an artificial man-made lake is created behind the dam.

Electricity is produced by a device called a turbine. Turbines contain metal coils surrounded by magnets. When the magnets spin over the metal coils, electricity is produced. Turbines are located inside dams. The falling water spins the magnets.

Dams provide clean, pollution free energy, but they can also harm the environment. Species that use rivers to spawn are often hurt by dams. In the Northwest, sockeye salmon and trout populations have dropped from 16 million to 2.5 million since hydroelectric plants were built on the Columbia River. Dams all over the world have hurt some species.

Types of Dams

(*i*) Based on their functions, dams can be classified as
- (*a*) Storage dams
- (*b*) Diversion dams
- (*c*) Detention dams

(*ii*) Based on the shape, the dams are classified into
- (*a*) Trapezoidal dams
- (*b*) Arch dams

(*iii*) Based on the materials of construction
- (*a*) Earth dams
- (*b*) Rock pieces dam
- (*c*) Stone masonary dams
- (*d*) Concrete dams
- (*e*) RCC dams
- (*f*) Timber and rubber dam

(*iv*) Based on hydraulic design
- (*a*) Overflow type dam
- (*b*) Non-overflow type dam

(*v*) Based on structural design
- (*a*) Gravity dam
- (*b*) Arch dam
- (*c*) Buttresses dam

2.5 SELECTION OF WATER TURBINES

The major problem confronting the engineering is to select the type of turbine which will give maximum economy. The hydraulic prime-over is always selected to match the specific conditions under which it has to operate and attain maximum possible efficiency.

The choice of a suitable hydraulic prime-mover depends upon various considerations for the given head and discharge at a particular site of the power plant. The type of the turbine can be determined if the head available, power to be developed and speed at which it has to run are known to the engineer beforehand.

The following factors have the bearing on the selection of the right type of hydraulic turbine which will be discussed separately:

1. Rotational Speed
2. Specific Speed
3. Maximum Efficiency
4. Part load Efficiency
5. Head
6. Type of Water
7. Runaway Speed
8. Cavitation
9. Number of Units
10. Overall cost

The hydraulic power of a plant can be calculated using following expression

$$P = g\,\rho\,QH$$

where, P is the hydraulic energy in Watts

g is acceleration due to gravity (9.81 m/s^2)

ρ is water density (1000 kg/m^3)

Q is the flow (or) discharge m^3/s

H is the height of fall of water or head in meter

The electric power produced in kWh.

$$W = 9.81 \times 1000 \times Q \times H \times \eta \times t$$
$$= 9.81\,Q\,H\,\eta\,t$$

where, t is the operating time in hours (8760 hours/year)

η is the efficiency of the turbine-generator (0.5 to 0.9)

The power developed depends upon the quantity of water (Q) and the head (H) of water.

Advantages of Hydropower plant

1. Water source is perennially available.
2. Running cost is very low.
3. Non-polluting.
4. Power generation can be switched on and off in a very short time.
5. Simple in concept and self-contained in operation.
6. Greater reliability.
7. Greater life (more than 50 years).
8. Other than power generation, it provides irrigation, flood control, afforestation navigation and aqua-culture.
9. High efficiency.
10. Suitable for spinning reserve.
11. Man power requirement is low.
12. Simple in design and operation.

Disadvantages of Hydropower

1. High capital investment and low rate of return.
2. Gestation period is very large.
3. Power generation depends on availability of water.
4. Plants are for away from load centre hence cost of transmission and losses are high.
5. Large hydro plants disturb the ecology of the area due to deforestation.

2.6 FACTORS TO BE CONSIDERED WHILE SELECTING THE SITE FOR HYDROELECTRIC POWER PLANT

1. Availability of water and water head.
2. Accessibility of site.
3. Water storage capacity.
4. Distance from the load centre.
5. Type of land.

2.7 PUMPED-STORAGE PLANTS

Pumped storage systems work using two reservoirs that are built at differing heights. During periods of peak demand, energy is generated by releasing water from the upper reservoir to drive turbines. The water is pumped back up to the upper reservoir from the lower reservoir during periods of low energy demand.

A pumped-storage plant has two reservoirs:

Upper reservoir: Like a conventional hydropower plant, a dam creates a reservoir. The water in this reservoir flows through the hydropower plant to create electricity.

Lower reservoir: Water exiting the hydropower plant flows into a lower reservoir rather than re-entering the river and flowing downstream.

Using a **reversible turbine**, the plant can pump water back to the upper reservoir. This is done in off-peak hours. Essentially, the second reservoir refills the upper reservoir. By pumping water back to the upper reservoir, the plant has more water to generate electricity during periods of peak consumption.

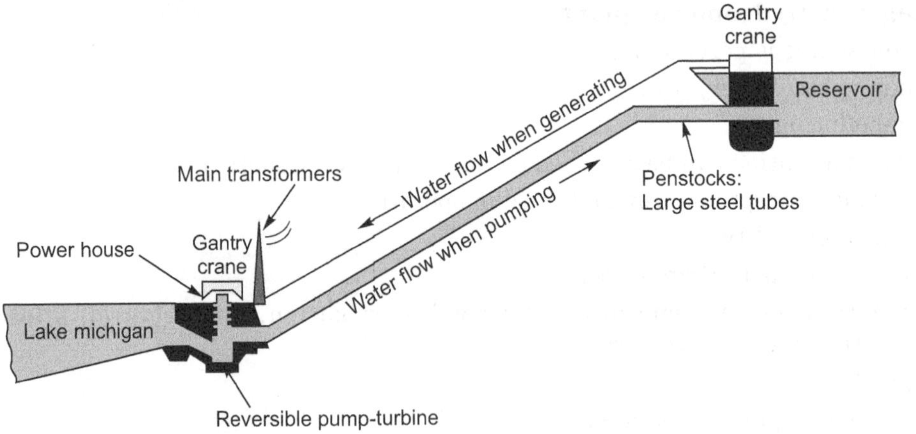

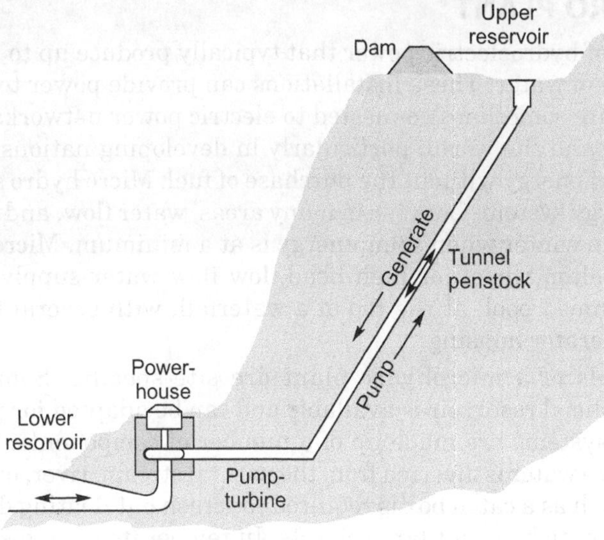

Fig. 2.2. Pumped Storage plant

The advantages of pumped storage hydro power plant are as follows:

(*i*) Increases the peak load capacity at low cost.

(*ii*) High operating efficiency.

(*iii*) Better load factor.

(*iv*) Independence of steam flow conditions.

2.8 MINI HYDRO PLANTS

The present energy crisis which is engulfing the world today is not showing any sign of abating. India is also gripped in this crisis as mentioned earlier. To develop mini (5 m to 20 m head) and micro (≪ 5 m) hydel potential in our country is partly a solution to face the present power crisis. The low head hydro-potential is scattered in this country and estimated potential from such sites could be as much as 20,000 MW.

In India, the high head and medium head sites have been explored but low head, low output sites are neglected. There is tendency to utilize high head sites because higher the head, lesser is the capital cost per kW. As high head plant provides higher velocity which reduces the size of turbine-generator considerably.

These sites are available at sluices admitting water to the canals for irrigation purposes and on the canal falls. At sluice inlets and canal falls, lot of energy is wasted because it cannot be utilized and costly civil engineering works have to be undertaken to dissipate this energy which would otherwise do great harm. To overcome this difficulty with an advantage of getting more power is to set up a small hydropower house which forms a small fraction of total cost.

If properly planned and implemented, it is possible to commission a small hydrogenerating set-up of 5 MW with a period of one and half year against the period of a decade or two for large capacity power plants.

A suitable type of turbine known as bulb turbine is developed for such small hydro plants. In this turbine, the scroll case has been modified in the form of nearly horizontal tapering duct with the flow passage forming a sort of throat where the turbine runner is located. A straight diverging tube acts as a draft tube. Thus the whole path of the water is straight resulting in minimum hydraulic losses.

2.9 MICRO HYDRO PLANT

Micro hydro is a type of hydroelectric power that typically produce up to 100 kW of electricity using the natural flow of water. These installations can provide power to an isolated home or small community, or are sometimes connected to electric power networks. There are many of these installations around the world, particularly in developing nations as they can provide an economical source of energy without the purchase of fuel. Micro hydro systems complement photovoltaic solar energy systems because in many areas, water flow, and thus available hydro power, is highest in the winter when solar energy is at a minimum. Micro hydro is frequently accomplished with apelton wheel for high head, low flow water supply. The installation is often just a small dammed pool, at the top of a waterfall, with several hundred feet of pipe leading to a small generator housing.

Construction details of a microhydro plant are site-specific. Sometimes an existing mill-pond or other artificial reservoir is available and can be adapted for power production. In general, micro hydro systems are made up of a number of components. The most important include the intake where water is diverted from the natural stream, river, or perhaps a waterfall. An intake structure such as a catch box is required to screen out floating debris and fish, using a screen or array of bars to keep out large objects. In temperate climates this structure must resist ice as well. The intake may have a gate to allow the system to be dewatered for inspection and maintenance.

The intake then tunnels water through a pipeline (penstock) to the powerhouse building containing a turbine. In mountainous areas, access to the route of the penstock may provide considerable challenges. If the water source and turbine are far apart, the construction of the penstock may be the largest part of the costs of construction. At the turbine, a controlling valve is installed to regulate the flow and the speed of the turbine. The turbine converts the flow and pressure of the water to mechanical energy; the water emerging from the turbine returns to the natural watercourse along a tailrace channel. The turbine turns a generator, which is then connected to electrical loads; this might be directly connected to the power system of a single building in very small installations, or may be connected to a community distribution system for several homes or buildings.

Usually micro hydro installations do not have a dam and reservoir, like large hydroelectric plants have, relying on a minimal flow of water to be available year-round.

Benefits:

- 'Fuel-free' source of power.
- Different to large hydro since environmental impacts of installation are negligible.
- Renewable energy source therefore helping to reduce greenhouse gas emissions and having a net positive impact on the environment.
- Constant generation over long periods unlike wind and solar power.
- Good correlation with demand (more hydro energy is available in winter when heating loads are high).
- Long lifetime of systems, typically 25 years or more Low maintenance requirements and running costs.
- Reasonable payback for grid-connected systems, often 10 years or less.

Merits of hydroelectric power plant

- Requires no fuels and hence pollution free
- Low operating cost and no need for specialized human resources for day-to-day operation

- Simple construction and requires less maintenance
- Very robust and durable
- The reservoir and dam used for power generation can be used for regulating flood and irrigation also
- Can also be used as base load power plant

Demerits of hydroelectric power plant

- Very high capital cost involved for dam construction
- Uncertainty about availability of huge quantity of water
- Skilled personnel required for construction
- High cost of transmission as plant is normally located at far off hilly areas construction of new hydroelectric power plant may necessitate rehabilitation of people and payment of compensation for land acquisition
- Period of construction of dam delays the early commissioning of the plant

2.10 PUMPED STORAGE HYDROELECTRIC POWER PLANTS IN INDIA

- Bhira, Maharashtra, 150 MW
- Kadamparai, Coimbatore, Tamil Nadu, 400 MW (4 × 100 MW)
- Nagarjuna Sagar PH, Andhra Pradesh, 810 MW (1 × 110 MW + 7 × 100 MW)
- Purulia Pumped Storage Project, Ayodhya Hills, Purulia, West Bengal, 900 MW
- Srisailam Left Bank PH, Andhra Pradesh, 900 MW (6 × 150 MW)
- Tehri Dam, Uttaranchal, 1,000 MW.

REVIEW QUESTIONS

Two Marks Questions

1. What is meant by hydroelectric power plant?
2. How the hydro power plant is classified?
3. What do you understand by headrace and tailrace?
4. How to do you estimate the power generators by hydroelectric power plant?
5. What do you meant by water hammer?
6. What is the purpose of surge tank in HEPP?
7. Give the advantage of hydel power plant.
8. Give the essential factors which should be considered while selecting site for hydel power plant.
9. What is penstock?
10. What are the advantages of hydro power plant?
11. What are the parameters to be considered to select the site for hydro power plant?
12. Draw the layout of hydro power plant?
13. How dams are classified?
14. What are the parameters to be considered for the selection of right type of hydraulic turbine?
15. What is hydraulic turbine? Give its classification.
16. What is impulse turbine and reaction turbine? Give an example for each.
17. Differentiate is impulse turbine and reaction turbine.
18. What is radial flow, axial flow and mixed flow turbine?
19. Differentiate francis turbine and Kaplan turbine.

20. Differentiate inward flow and outward flow reaction turbine.
21. What is draft tube? Explain its function.
22. Define unit speed, unit discharge and unit power.
23. Define specific speed of a turbine.
24. Classify turbines for micro hydel power plant.
25. Draw electrical block diagram of battery based micro hydro system.
26. What are the advantages of micro hydro plant?
27. What are the different types of energy storage system?
28. Classify hydro by size.
29. What is the function of guide vanes in francis turbine?
30. What is meant by pumped storage plants?
31. What is mini and micro hydro plant?
32. What is the function of hydro turbine?
33. How hydro turbines are classified?
34. What are the benefits of hydro plant?
35. List out any two merits and demerits of hydroelectric power plant.
36. List out any two pumped storages hydroelectric power plant in India.
37. Give the load curve for the pumped storage system.
38. What is forebay?
39. What are the main parts of pelton wheel?
40. What are the different types of spillways?
41. What is speed rings or stay rings?
42. What are the different operating cycles in pumped storage plants?

Sixteen Marks Questions

1. Explain the working principle of hydroelectric power plant with a neat layout.
2. Describe the following:
   (*i*) Dams (*ii*) Hydro turbines
3. Discuss the factors to be considered in selecting the water turbine for hydroelectric power plant.
4. How are dams classified? What are the factors to be considered in selecting type of dams?
5. What is meant by pumped storage plant? Explain with a neat layout.
6. Explain pumped storage plant advantages and disadvantages.
7. Explain with neat sketch the function of surge tank in hydroelectric power plant.
8. Neatly explain the hydro electric energy resources in India.
9. Explain:
   (*i*) Mini hydro plant (*ii*) Micro hydro plant.
10. List out the merits and demerits of hydroelectric power plant.

3 Nuclear Power Plant

3.1 PRINCIPLES OF NUCLEAR ENERGY

Nuclear energy has the potential to be a reliable, sustainable and environmentally friendly energy source that can contribute to the accessibility of affordable energy services in all interested countries for present and future generations. Any use of nuclear energy should be beneficial, responsible and sustainable, with due regard to the protection of people and the environment, non-proliferation, and security.

Beneficial Use

Benefits: The use of nuclear energy should provide benefits that outweigh the associated costs and risks.

Transparency: The use of nuclear energy should be based on open and transparent communication of all its facets.

Responsible Use

Protection of people and the environment: The use of nuclear energy should be such that people and the environment are protected in compliance with the IAEA Safety Standards and other internationally recognized standards.

Security: The use of nuclear energy should take due account of the risk of the malicious use of nuclear and other radioactive material.

Non-proliferation: The use of nuclear energy should take due account of the risk of the proliferation of nuclear weapons.

Long-term commitment: The use of nuclear energy should be based on a long-term commitment.

Sustainable Use

Resource efficiency: The use of nuclear energy should be efficient in using resources.

Continual improvement: The use of nuclear energy should be such that it pursues advances in technology and engineering to continually improve safety, security, economics, proliferation resistance, and protection of the environment.

3.2 ELEMENTS AND ATOMS

An element is a pure chemical substance that has one or one type of atom, distinguished by its atomic number. The atomic number is derived from the number of protons present in the element's nucleus. There are a total of 118 elements that have been identified, divided between metal, metalloids and non-metals. Each element has its own set of properties. Most elements are available on the earth, while a few have been developed artificially through nuclear reactions.

Atoms are the basic units that all matter is made of Atoms are tiny, ranging from 0.1 to 0.5 nanometers in width. They are so tiny that they cannot even be seen through a microscope. There are many types of atoms. Each atom has a unique name, mass and size. The different types of atoms are called elements.

Essentially, atom is a unit whereas elements are specific substances. An atom is the smallest possible units of that element that retains the properties of that element. For example: Iron (Fe) is an element. A block of iron can be broken down into iron flakes, which can then be further broken down into atoms. The atom would still exhibit all the properties of iron including magnetism.

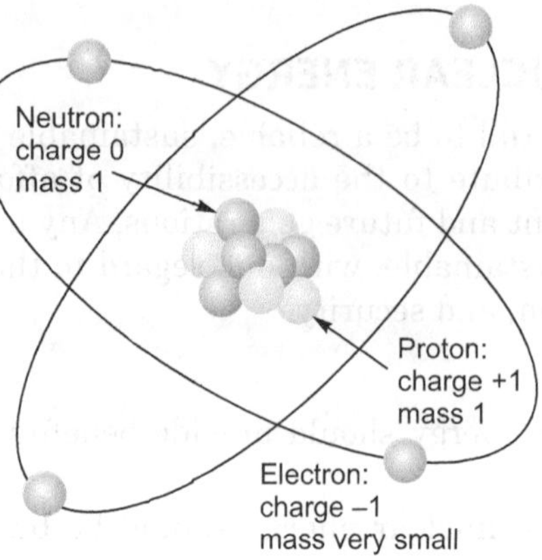

Fig. 3.1. Elements and Atoms

Each atom of any element is made up of protons, neutrons and electrons. The protons and neutrons make up the nucleus of the atom and are situated in the middle of the atom. The nucleus is surrounded by a cloud of electrons which are bound to the nucleus by an electromagnetic force. The electrons have a negative charge which is how they are attracted to the nucleus, as the protons in the nucleus have a positive charge. Neutrons, on the other hand, do not have a charge.

The number of protons, neutrons and electrons in an atom determine which element it is. For example: An atom of iron has 26 protons, 30 neutrons and 26 electrons. Each atom of iron will have this configuration.

Elements are the basic building blocks for all types of substances. They can combine with each other to form molecules. Atoms of different elements come together to make molecules. This happens via a chemical reaction. For example: two hydrogen atoms and one oxygen atom combine to make a water molecule.

3.3 RADIOACTIVITY

Radioactivity is a 1975 concept album by Kraftwerk. It was also released under the German name of Radio-Aktivitat. Radioactivity is the spontaneous disintegration of atomic nuclei.

The nucleus emits α particles. β particles, or electromagnetic rays during this process. It is an important source of energy for small power devices and a source of radiation for use in research, industry, medicine and a wide variety of applications.

Units of Radioactivity

- **Roentgen (*R*):** Defined as the amount of ionizing radiation which produces 2.08×10^9 ion pairs in 1 cm^3 of air.
- **Radiation Absorbed Dose (RAD):** A rad is the amount of radiation that puts 10 < SUP – 2 < SUP > *J* of energy in 1 kg of absorbing material.
- **Relative Biological Effectiveness (RBE):** The biological risk *a*, *B*, and *Y* radiation differ; The RBE factor compares the number of rads of *X*-radiation or *Y*-radiation that produce the same biological damage as a rad of the radiation used.
- **Roentgen Equivalent in Man (REM):** Product of amount of rad and the RBE factor.
- **Gray (*Gy*)**—100 rads.
- **Sievert (*Sv*)**—100 rem.

3.4 BIOLOGICAL EFFECTS OF RADIATION

- Ionizing radiation causes physical damage to cells and DNA.
- At high doses of radiation (10,000–15,000 rads), death occurs in a few hours because of neurological and cardiovascular breakdown (Central Nervous Syndrome).
- Medium doses, 500–1200 rads, cause death to occur in a few days because of the destruction of the gastrointestinal mucosa.
- Lower doses, 250–500 rads, causes death to occur after several weeks due to damage of the blood forming organs (hematopoietic syndrome).

3.5 RADIOACTIVE DECAY

Atoms are continually undergoing decay. When studying nuclear chemistry, there is a typical format used to represent specific isotopes.

Mass number

↓

$^{238}_{92}U$ ⟵ Element

↑

Charge (or) Atomic number

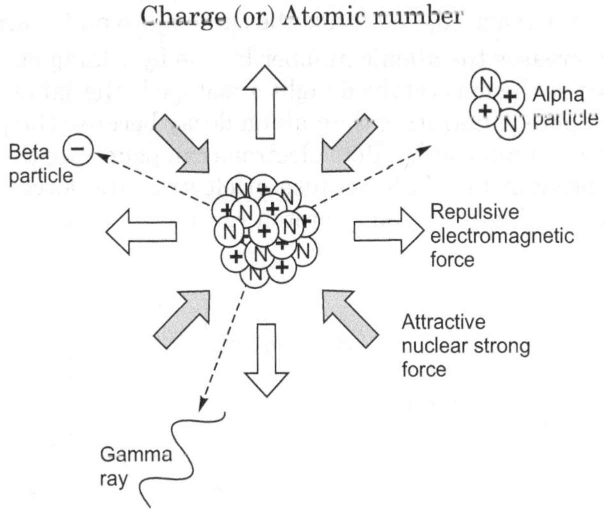

Fig. 3.2. Radioactive decay

There are five different types of radioactive decay:

- Alpha decay (α)
- Beta negative decay (β)
- Gamma decay (γ)
- Positron emission (Beta positive decay) (β^+)
- Electron capture

1. Alpha decay

$$ {}^{x}_{z}A \longrightarrow {}^{4}_{2}He \longrightarrow {}^{z-4}_{z-2}B $$

Alpha article ←

where *A* is the parent isotope, *B* is the daughter isotope or the isotope formed When an element is broken down in alpha decay it looses two neutrons and two protons, (Figure 3.2) this means that the name of the element will change as well, moving back two places on the periodic table. Alpha decay is not very penetrating because the He atoms capture electrons before traveling very far.

However, it is very damaging because the alpha particles can knock atoms off of molecules. Alpha decay is the most common in elements with an atomic number greater than 83. Its main radiation hazard comes when it is ingested into the body; it has great destructive power within its short range. In contact with fast-growing membranes and living cells, it is positioned for maximum damage.

$$ Pu^{239} \longrightarrow U^{235} + \alpha \text{ particle (He–4 nucleus)} $$

2. Beta negative decay

$$ {}^{x}_{z}A \longrightarrow {}^{0}_{-1}e^{-1} \longrightarrow {}^{x}_{z+1}B $$

Beta particle ←

There are two types of β decay: β+ and β– decay. An excess of neutrons in an atom's nucleus will make it unstable, and a neutron is converted into a proton to change this ratio. During this process, a β particle is released, and it has the same mass and charge as an electron.

The beta emission increases the atomic number by one by adding one proton. At the same time, one neutron is lost so the mass of the daughter isotope is the same as the parent isotope. Beta negative decay is more penetrating than alpha decay because the particles are smaller, but less penetrating than gamma decay. Beta electrons can penetrate through about one cm of flesh before they are brought to a halt because of electrostatic forces. Beta decay is most common in elements with a high neutron to proton ratio.

3. Gamma decay

$$ {}^{x}_{z}A \longrightarrow {}^{0}_{0}\gamma \longrightarrow {}^{x}_{z}B $$

Gamma particle ←

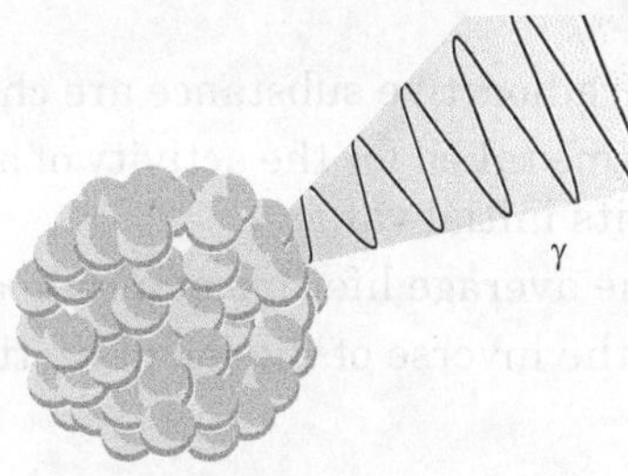

Gamma Ray

Gamma ray emission usually occurs with and β emission. Gamma rays have no charge or mass, so their emission doesn't change the chemical composition of the atom. Instead, it results in a loss of radiant energy. Gamma ray emission occurs because the nucleus is often unstable after α and β decay. Gamma radioactivity is composed of electromagnetic rays.

It is distinguished from *X*-rays only by the fact that it comes from the nucleus. Most gamma rays are somewhat higher in energy than *X*-rays and therefore are very penetrating. It is the most useful type of radiation for medical purposes, but at the same time it is the most dangerous because of its ability to penetrate large thicknesses of material.

4. Positron emission (also called Beta positive decay)

In this reaction a positron is emitted. A positron is exactly like an electron in mass and charge force except with a positive charge.

$${}^{x}_{z}A \longrightarrow {}^{0}_{0}e \longrightarrow {}^{x}_{z-1}B$$

Positron particle ←

It is formed when a proton breaks into a neutron with mass and neutral charge and this positron with no mass and the positive charge. Positron emission is most common in lighter elements with a low neutron to proton ratio.

5. Electron capture

$${}^{x}_{z}A \longrightarrow {}^{0}_{-1}e \longrightarrow {}^{x}_{z-1}B$$

Electron particle ←

In this reaction a nucleus captures one of its own atom's inner shell electrons which reduces the atomic number by one. This captured electron joins with a proton in the nucleus to form a neutron. Electron capture is common in larger elements with a low neutron to proton ratio.

All elements with an atomic number over 83 are considered radioactive.

Decay Chains

A radioactive decay series is the chain of decays that occur starting with a radioactive isotope. An example of this is the uranium-radium series:

Uranium-238 α decays ⟶ thorium-234

Thorium-234 decays ⟶ protactinium-234

Protactinium-234 β decays ⟶ uranium-234

Uranium-234 α decays ⟶ thorium-230

Thorium α decays ⟶ radium-226

Radioactive Decay rate

The decay rate, or activity, of a radioactive substance are characterized by:

- ***Half-life*** – symbol $t_{1/2}$—the time taken for the activity of a given amount of a radioactive substance to decay to half of its initial value
- Mean lifetime – symbol τ—the average lifetime of a radioactive particle
- Decay constant – symbol λ—the inverse of the mean lifetime.

Half-life

All radioactive elements disintegrate according to their specific half-life. The half-life of a radioactive substance is the time required for half of the initial number of nuclei to disintegrate. The decay rate expresses the speed at which a substance disintegrates. It is the time taken for half the radionuclide's atoms to decay.

The rate of radioactive decay is typically expressed in terms of either the radioactive half-life, or the radioactive decay constant. They are related as follows:

$$T_{1/2} = \frac{\text{In}2}{\lambda} \approx \frac{0.693}{\lambda} \approx 0.693\tau$$

Radioactive half-life | Radioactive Decay constant | Mean lifetime

The decay constant is also sometime called the disintegration constant. The half-life and the decay constant give the same information, so either may be used to characterize decay. Another useful concept in radioactive decay is the average lifetime. The average lifetime is the reciprocal of the decay constant as defined here.

Decay timing

The number of decay events – dN expected to occur in a small interval of time dt is proportional to the number of atoms present. If N is the number of atoms, then the probability of decay $(-dN/N)$ is proportional to dt

$$\left[\frac{-dN}{N}\right] = \lambda\, dt$$

Particular radionuclides decay at different rates, each having its own decay constant (A). The negative sign indicates that N decreases with each decay event. The solution to this first-order differential equation is the following function

$$N(t) = N_0 e^{-\lambda} = N_0 e^{-t/\tau}$$

where N_0 is the value of N at time zero (t = 0). The second equation recognizes that the differential decay constant λ has units of 1/time, and can thus also be represented as $1/\tau$, where τ is a characteristic time for the process. This characteristic time is called the time constant of the process. In radioactive decay, this process time constant is also the mean lifetime for decaying atoms. Each atom "lives" for a finite amount of time before it decays, and it may be shown that this mean lifetime is the arithmetic mean of all the atoms' lifetimes and that it is t, which again is related to the decay constant as follows:

$$\tau = \frac{1}{\lambda}$$

This relationship between the half-life and the decay constant shows that highly radioactive substances are quickly spent, while those that radiate weakly endure.

The factor of ln2 in the above relations results from the fact that concept of "half-life" is merely a way of selecting a different base other than the natural base e for the lifetime expression. The time constant t is the "1/e" life rather than the "1/2" life of radionuclide.

The result from the decay probability can be put in the differential form:

$$dN = -\lambda N dt \quad \text{or} \quad \frac{dN}{N} = -\lambda\, dt$$

This can be integrated directly to give In $N = -\lambda t + C$ where C is a constant of integration. Taking the exponent of both sides gives

$$N = e^c e^{-\lambda t} \text{ and } N = e^c = N_0 \text{ at } t = 0$$

so the standard form of the decay equation is:

$$N = N_0 e^{-\lambda t}$$

which also applies to all other quantities which are proportional to N.

Electron capture (*K*-capture)

In all the cases where β+ decay is allowed energetically, it is accompanied by the electron capture process, when an atomic electron is captured by a nucleus with the emission of a neutrino (small neutral one) denoted by the Greek word (v_e).

$$\text{energy} + p + e^- \longrightarrow n + v_e$$

But if the energy difference between initial and final states is less than $2m_ec^2$, then β+ decay is not energetically possible, and electron capture is the sole decay mode. This decay is also called *K*-capture, because the 'inner most' electron of an atom belongs to the *K*-shell of the electronic configuration of the atom and this has the highest probability to interact with the nucleus.

3.6 NUCLEAR LIFE CYCLE

The Nuclear Fuel Cycle begins when uranium is mined, enriched, and manufactured into nuclear fuel, which is delivered to a nuclear power plant. (Figure 3.3). After usage in the power plant, the spent fuel is delivered to a reprocessing plant or to a final repository for geological disposition. In reprocessing 95% of spent fuel can be recycled to be returned to usage in a power plant.

A nuclear reactor is only part of the life-cycle for nuclear power. The process starts with mining. Uranium is a slightly radioactive material which is found at low concentrations throughout the world. The average concentration in the earth's crust is about 2.7 parts per million.

The type of mine developed depends on the nature of the deposit. Uranium mines are underground, surface or open-pit, or in-situ leach mines. The underground mining has to be undertaken where the deposits are more than 100 m below the surface. Underground mines requires careful ventilation to protect miners against radiation exposure.

The process of milling extracts Uranium from the Ore produced in the mine. In any case, the uranium ore is extracted, usually converted into a stable and compact form such as Yellow cake, and then transported to a processing facility. The Yellow cake typically contains more than 60% of Uranium.

Purification and conversion is needed since yellow cake is not directly usable as nuclear reactor fuel.

Natural Uranium consists mainly two forms. It is termed as isotopes Uranium-235 and Uranium-238. Uranium-235 is the most important isotopes since it undergoes fission much more readily than Uranium-238 in the nuclear reactor.

At this point, the enriched uranium, containing more than the natural 0.7% U-235. In most types of reactor, a higher concentration of uranium is used to make Fuel rod. This produced by a process termed enrichment.

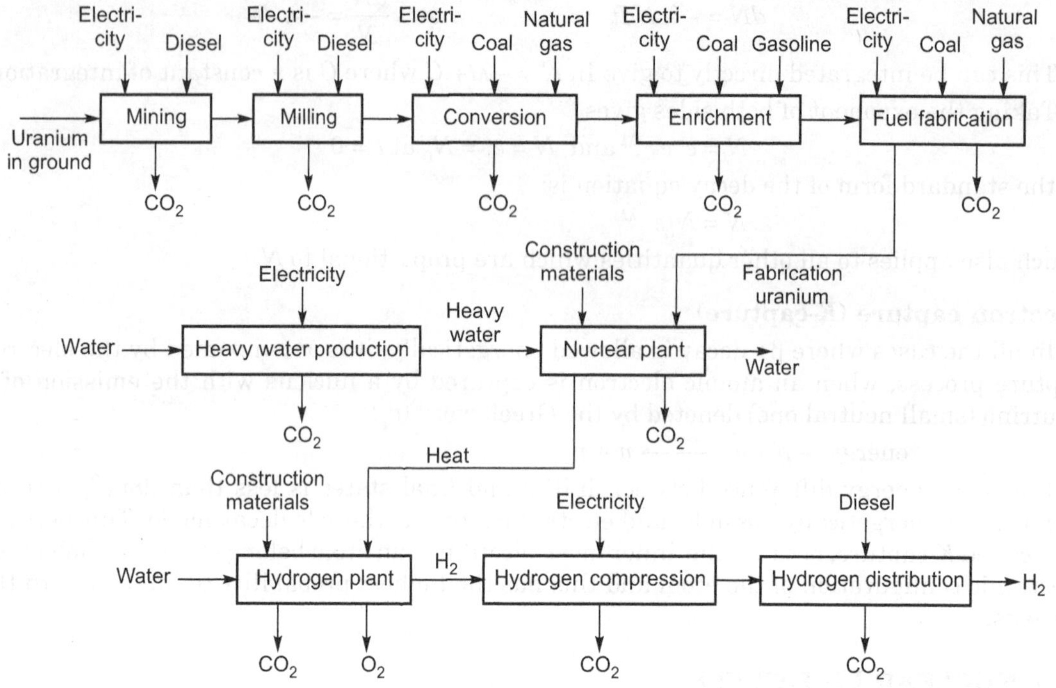

Fig. 3.3. Nuclear Life Cycle

Uranium enrichment can be done by two ways.

1. Gas centrifuge process
2. Gas diffusion

Here the Uranium-235 concentration is increased to 2–4%.

The Enriched Uranium hexafluoride is converted to enriched Uranium Oxide as the input for fuel production. Fuel loaded into nuclear reactors undergoes fission of the Uranium 235 atoms and this process releases energy. This energy is used to heat water and produce steam which drives a turbine. The turbine in turn drives a Generator producing electricity which is distributed by the electricity grid system.

During operation of the reactors, a proportion of the uranium atoms is transformed into other elements by fission of by absorption of neutrons. These elements include fission products which are radioactive wastes and Plutonium.

The used fuel contains 96% uranium, 1% plutonium and 3% radioactive wastes. Reprocessing is used to separate the waste from the uranium and plutonium which can then be recycled into new fuel. The reprocessing effectively reduces the volume of waste and limits the need to mine new supplies of uranium, so that extending the life time of the resources.

The production of radioactive waste is minimized as far as reasonably practicable and the reuse recycling of materials are encouraged.

Radioactive wastes from the nuclear fuel cycle are categorized as high, medium or low level according to the intensity of radiation they emit. The low level waste is produced at all stages of the nuclear fuel cycle, intermediate waste arises mainly during reactor operations and reprocessing and high level waste comprises spent fuel and waste containing fission products from reprocessing. All radioactive wastes have to be carefully managed to ensure high levels of safety are maintained.

3.7 NUCLEAR ENERGY—FISSION AND FUSION

One of the laws of the universe is that matter and energy can't be created nor destroyed. But, they can be changed in form. Matter can be changed into energy. The world's most famous scientist, Albert Einstein, created the mathematical formula that explains this

$$E = mc^2$$

where E = energy, m = mass, c = stands for the velocity or the speed of light.

Scientists used Einstein's famous equation as the key to unlock atomic energy and also create atomic bombs.

Nuclear fission

An atom's nucleus can be split apart. When this is done, a tremendous amount of energy is released. The energy is both heat and light energy. Einstein said that a very small amount of matter contains a very amount of energy. This energy, when let out slowly, can be harnessed to generate electricity. When it is let out all at once, it can make a tremendous explosion in an atomic bomb.

A nuclear fission power plant uses the heat generated by a nuclear fission process to drive a steam turbine which generates usable electricity. The underlying physical process is the same for all power plant designs, what differs is the way in which the nuclear reaction is controlled.

This is an engineering problem which must take into account factors such as:

- *Control*—Keeping the nuclear reaction from dying out or exploding.
- *Safety*—If something goes wrong it can be contained.
- *Refueling*—Adding more nuclear fuel without stopping the reactor.
- *Waste production*—The byproducts of the reaction must be manageable.
- *Efficiency*—Capture as much of the heat as possible.

A nuclear power plant uses uranium as a fuel. Uranium is an element that is dugout of the ground many places around the world. It is processed into tiny pellets that are loaded into very long rods that are put into the power plant's reactor. The word fission means to split apart.

Inside the reactor of an atomic power plant, uranium atoms are split apart in a controlled chain reaction. Uranium—which constantly undergoes spontaneous fission very slowly. This is why the element emits radiation, and why it's a natural choice for the induced fission that nuclear power plants require.Uranium is a common element on Earth. Uranium-235 has an interesting property that makes it useful for both nuclear power production and for nuclear bomb production. U-235 decays naturally, just as U-238 does, by alpha radiation. U-235 also undergoes spontaneous fission a small percentage of the time. However U-235 is one of the few materials that can undergo induced fission. If a free neutron runs into a U-235 nucleus, the nucleus will absorb the neutron without hesitation, become unstable and split immediately.

Something on the order of 20 MeV (million electron volts) is released by the decay of one U-235 atom (1 eV is equal to 1.602×10^{-12} ergs, 1×10^7 ergs is equal to 1 joule, 1 joule equals 1 watt-second, and 1 BTU equals 1,055 joules).

The probability of a U-235 atom capturing a neutron as it passes by is high. In fact under reactor conditions, one neutron ejected from each fission causes fission to occur. As soon as the nucleus captures the neutron, it splits into two lighter atoms and throws of two or three new neutrons (the number of ejected neutrons depends on how the U-235 atom splits). The process of capturing the neutron and splitting happens very quickly, on the order of picoseconds (1×10^{-12} seconds).

$$^{235}\text{U} + \text{neutron} \longrightarrow \text{fission fragments} + 2.4 \text{ neutrons} + 192.9 \text{ MeV}$$

$$^{239}\text{Pu} + \text{neutron} \longrightarrow \text{fission fragments} + 2.9 \text{ neutrons} + 198.5 \text{ MeV}$$

Note that these equations are for fissions caused by slow-moving (thermal) neutrons. The average energy released and number of neutrons ejected is a function of the incident neutron speed. The splitting of an atom releases an incredible amount of heat and gamma radiation, or radiation made of high-energy photons (Figure 3.4). The two atoms that result from the fission later release beta radiation (super fast electrons) and gamma radiation of their own as well. The energy released by a single fission comes from the fact that the fission products and the neutrons, together, weigh less than the original U-235 atom.

The difference in weight is converted directly to energy at a rate governed by the equation $E = mc^2$. Three-percent enrichment is sufficient for nuclear power plants, but weapons-grade uranium is composed of at least 9 per cent U-235. Two typical fission reactions are shown below with average values of energy released and number of neutrons ejected.

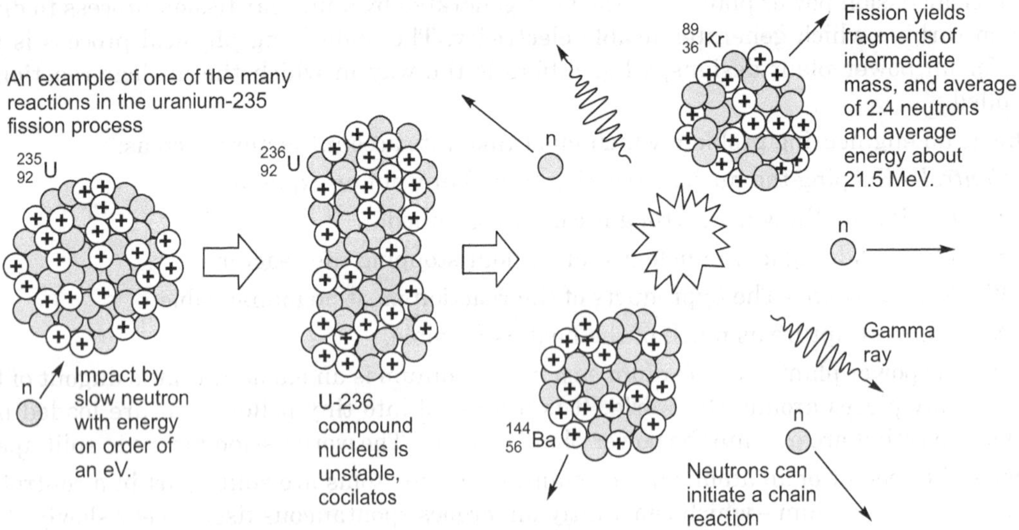

Fig. 3.4. Uranium-235 Fission

Neutron lifetime

The prompt neutron lifetime (l), is the average time between the emission of neutrons and either their absorption in the system or their escape from the system. The term lifetime is used because the emission of a neutron is often considered its birth, and the subsequent absorption is considered its death.

Slow-neutron fission reactors—order of 10^4 seconds and Fast fission reactors—order of 10^7 seconds. The extremely short lifetimes mean that in l second. 10,000 to 10,000,000 neutron lifetimes.

Mean Generation Time

It is the average time from a neutron emission to a capture that results in fission. The mean generation time is different from the prompt neutron lifetime because the mean generation time only includes neutron absorptions that lead to fission reactions.

$$A = (1/k)$$

k is the effective neutron multiplication factor.

The Four factor formula effective neutron multiplication factor, k is the average number of neutrons from one fission that cause another fission. The remaining neutrons either are absorbed in non-fission reactions or leave the system without being absorbed.

The value of k determines how a nuclear chain reaction proceeds.

Subcriticality k < 1

The system cannot sustain a chain reaction, and any beginning of a chain reaction dies out over time. For every fission that is induced in the system, an average total of $1/(1-k)$ fissions occur.

Criticality k = 1

Every fission causes an average of one more fission, leading to a fission level that is constant. Nuclear power plants operate with $k = 1$ unless the power level is being increased or decreased.

Supercriticality k > 1

For every fission in the material, it is likely that there will be k fissions after the next mean generation time. The result is that the number of fission reactions increases exponentially, according to the equation $e\ (k >)t$, where t is the elapsed time. Nuclear weapons are designed to operate under this state.

There are two subdivisions of supercriticality: prompt and delayed.

In an infinite medium, neutrons cannot leak out of the system and the multiplication factor becomes the infinite multiplication factor, which is approximated by the four-factor formula. The four factor formula is used in nuclear engineering to determine the multiplication of a nuclear chain reaction in an infinite medium. The formula is:

$$K_\infty = \eta\, fp\varepsilon$$

- η – Reproduction Factor—The number of fission neutrons produced per absorption in the fuel
- f – The thermal utilization factor—Probability that a neutron that gets absorbed does so in the fuel material
- p – The resonance escape probability—Fraction of fission neutrons that manage to slow-down from fission to thermal energies without being absorbed
- ε – The fast fission factor

Nuclear fusion

Another form of nuclear energy is called fusion. Fusion means joining smaller nuclei to make a larger nucleus (Figure 3.5). The sun uses nuclear fusion of hydrogen atoms into helium atoms. This gives off heat and light and other radiation.

Scientists have been working on controlling nuclear fusion for a long time, trying to make a fusion reactor to produce electricity. But, they have been having trouble learning how to control the reaction in a contained space.

What's better about nuclear fusion is that it creates less radioactive material than fission, and its supply of fuel can last longer than the sun.

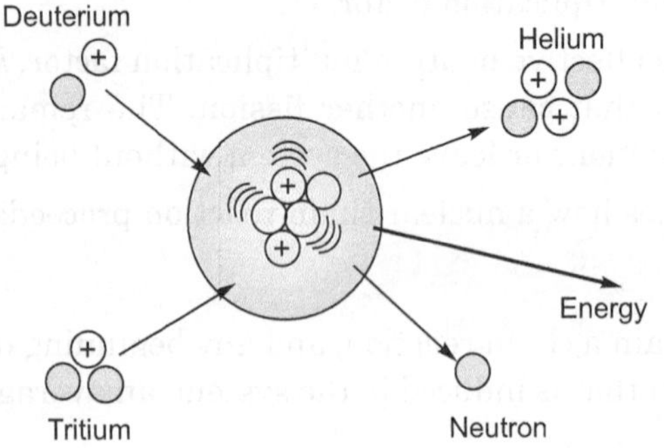

Fig. 3.5. Nuclear Fusion

Fusion chain reaction, it occurs under extreme pressure and temperature conditions. which are maintained by the energy released in the fusion process. Fusion of deuterium with tritium creating helium-4, freeing a neutron, and releasing 17. Electronvolt MeV of energy, as an appropriate amount of mass converting to the kinetic energy of the products in agreement with

$$E = mc^2$$

3.8 URANIUM-235 CHAIN REACTION

If an least one neutron from U-235 fission strikes another nucleus and causes it to fission, then the chain reaction will continue. If the reaction will sustain itself, it is said to be critical, and the mass of U-235 required to produce the critical condition is said to be a critical mass. A critical chain reaction can be achieved at low concentrations of U-235 if the neutrons from fission are moderated to lower their speed, since the probability for fission with slow neutrons is greater.

A fission chain reaction produces intermediate mass fragments which are highly radioactive and produce further energy by their radioactive decay. Some of them produce neutrons, called delayed neutrons, which contribute to the fission chain reaction.

In a chain reaction, particles released by the splitting of the atom go off and strike other uranium atoms splitting those (Figure 3.6). Those particles given off split still other atoms in a chain reaction. In nuclear power plants, control rods are used to keep the splitting regulated so it doesn't go too fast.

It's been around since the planet formed. Uranium-238 (U-238) has an extremely long half-life of 4.5 billion years. It makes up 9 percentage of the uranium on Earth, while uranium-235 (U-235) makes up about 0.7 percentage of the remaining uranium found naturally.

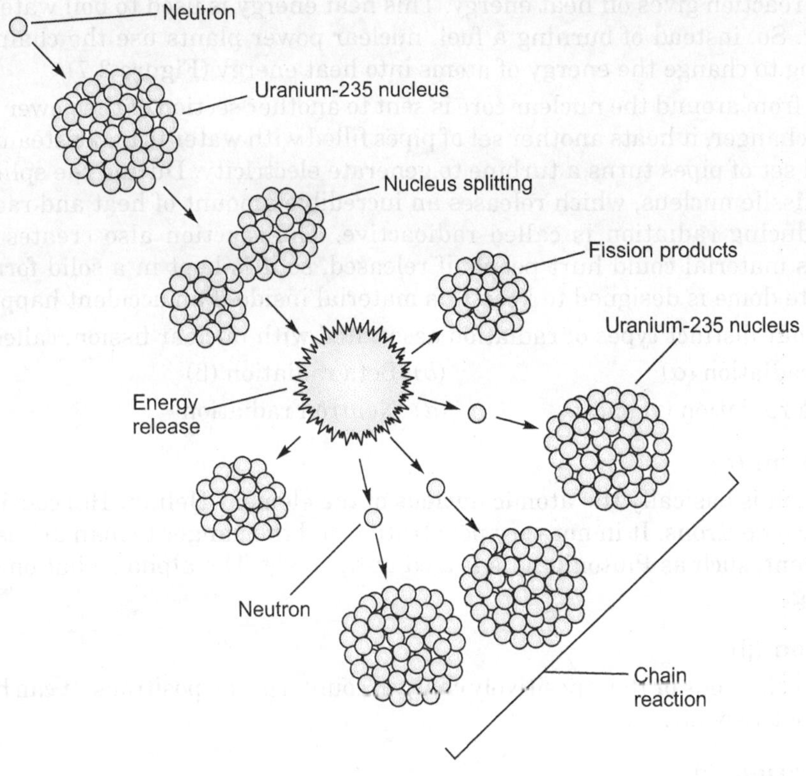

Fig. 3.6. Chain Reaction

Uranium-234 is even rarer, formed by the decay of U-238. U-238 goes through main stages of decay in its life span, eventually forming a stable isotope of lead, so U-234 is just one link in that chain.

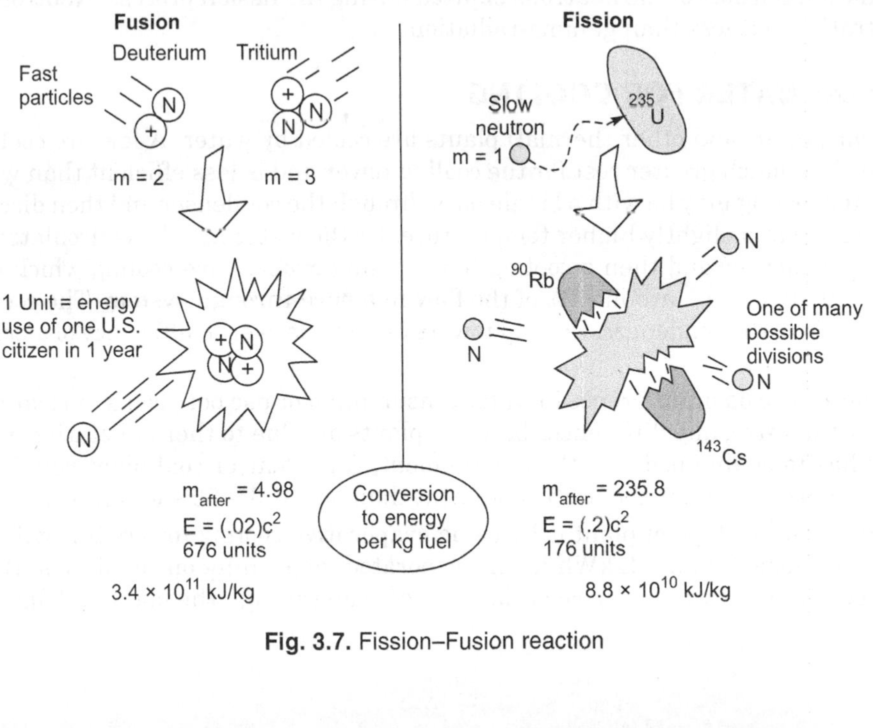

Fig. 3.7. Fission–Fusion reaction

This chain reaction gives off heat energy. This heat energy is used to boil water in the core of the reactor. So, instead of burning a fuel, nuclear power plants use the chain reaction of atoms splitting to change the energy of atoms into heat energy (Figure 3.7).

This water from around the nuclear core is sent to another section of the power plant. Here, in the heat exchanger, it heats another set of pipes filled with water to make steam. The steam in this second set of pipes turns a turbine to generate electricity. During the splitting of U^{235} or any other fissile nucleus, which releases an incredible amount of heat and radiation. Any material producing radiation is called radioactive. The reaction also creates radioactive material. This material could hurt people if released, so it is kept in a solid form. The very strong concrete dome is designed to keep this material inside if an accident happens.

There are four distinct types of radiation associated with nuclear fission, called

(*a*) Alpha radiation (α)
(*b*) Beta radiation (β)
(*c*) Gamma radiation (γ) and
(*d*) Neutron radiation

Alpha radiation (α)

This radiation is basically the atomic nucleus of the element Helium(He) consisting of two protons and two neutrons. It in not very penetrative and the danger to man arises if an alpha emitting element, such as Plutonium, is lodged in the body. The alpha-radiation can then be very damaging.

Beta radiation (β)

It consists of electrons or their positively charged counterparts, positrons. It can be penetrate the skin, but not very far.

Gamma radiation (γ)

It is penetrative in a manner similar to *X*-rays and has similar physical properties. It can be stopped only by thick shields of lead or concrete.

Neutron radiation

The radiation consists of the neutrons emitted during the fission process. Neutrons are also very penetrative, but less than gamma-radiation.

3.9 USE OF WATER FOR COOLING

Most nuclear power (and other thermal) plants are cooled by water. A few are cooled by air, but this involves much greater cost for the cooling tower and is less efficient than wet cooling towers. Water cooling may be with a single pass through the condenser and then discharged to river, lake or sea at a slightly higher temperature. Or the water may be recirculated, passing through the condenser and then a cooling tower, using evaporative cooling which consumes (evaporates) some water—around 5% of the flow in a once-through system. The cooled water then is returned to the condenser. Cooling towers reduce the overall efficiency of a power plant by 3–5%.

In fact there is no basic difference in water consumption or use between a nuclear and a coal plant. Apart from size, any differences between plants are due to thermal efficiency, *i.e.*, how much heat has to be dumped into the environment. A nuclear or coal plant running at 32% efficiency will need to dump about 5% more heat than one at 35–36% efficiency.

If cooling for either type of plant is by using evaporative cooling towers this will translate into water use, some 2 to 2.5 L/kWh being evaporated, depending on conditions. But, if it is using a large volume of water (big river, lake or sea) run through the plant and discharged at a slightly higher temperature, there is no use in the sense of depletion. Cooling water is not polluted by use. Most nuclear power plants are cooled in the latter fashion. Since their location is not at all determined by the fuel.

3.10 NUCLEAR POWER PLANT REACTORS

All nuclear reactors are devices designed to maintain a Chain reaction producing a steady flow of neutrons generated by the fission of heavy nuclei. In these reactors, the fission of heavy atomic nuclei, the most common of which is uranium-235, produces heat that is transferred to a fluid which acts as a coolant.

During the fission process, bond energy is released and this first becomes noticeable as the kinetic energy of the fission products generated and that of the neutrons being released. Since these particles undergo intense deceleration in the solid nuclear fuel, the kinetic energy turns into heat energy.

In the case of reactors designed to generate electricity, to which the explanations below will now be restricted, the heated fluid can be gas, water or a liquid metal. The heat stored by the fluid is then used either directly (in the case of gas) or indirectly (in the case of water and liquid metals) to generate steam. The heated gas or the steam is then fed into a turbine driving an alternator.

The Reactors are, however, differentiated either by their purpose or by their design features. In terms of purpose, they are either (*i*) Research reactors, (*ii*) Power reactors.

Research reactors are operated at universities and research centres in many countries, including some where no nuclear power reactors are operated. These reactors generate neutrons for multiple purposes, including producing radiopharmaceuticals for medical diagnosis and therapy, testing materials and conducting basic research.

Power reactors are usually found in nuclear power plants. Dedicated to generating heat mainly for electricity production, they are operated in more than 3 countries. Their lesser uses are drinking water or district water production. In the form of smaller units, they also power ships.

Breeder reactor

Breeder reactors are intended to produce nuclear fuels in bulk from more abundant isotopes. The better known Fast breeder reactor makes ^{239}Pu (a nuclear fuel) from the naturally very abundant ^{238}U (not a nuclear fuel). Differentiating nuclear reactors according to their design features is especially pertinent when referring to nuclear power reactors (*i.e.,* Types of Nuclear Power Reactors).

A nuclear reactor produces and controls the release of energy from splitting the atoms of certain elements. In a nuclear power reactor, the energy released is used as heat to make steam to generate electricity.

Reactors can be heterogeneous or homogeneous. A heterogeneous reactor has a large number of fuel rods with the coolant circulating around them and carrying away the heat released by nuclear fission. In other case, the fuel and moderator are mixed (*i.e.,* Fissionable salt of uranium like uranium sulphate dissolved in the moderator (H_2O or D_2O).

These reactors are classified based on:

- the type of fuel used
- the coolant and
- the neutron flux spectrum
- the moderator

The most widespread power plant reactor types are:

1. Pressurized Water Reactor (PWR)
2. Boling Water Reactor (BWR)
3. Pressurized Heavy Water Reactors (PHWR)
4. Liquid Metal Fast Breeder Reactors (LMFBR)
5. High Temperature Gas Cooled Reactors (HTGR)
6. Advanced Gas Cooled Reactors (AGR)
7. Magnox
8. Advanced Pressurized Water Reactor (APWR)
9. Advanced Liquid Metal Reactor (ALMR)
10. Advanced Boiling Water Reactor (ABWR)
11. Integral Fast Reactor (IFR)
12. Modular High Temperature Gas Cooled Reactor (MHTGR)
13. Simplified Boiling Water Reactor (SBWR)

Few of above reactors are explained in the following sections.

Pressurized Water Reactor (PWR)

Nuclear power plants run on uranium fuel. In the reactor, uranium atoms are split through a process known as fission. When atoms are spilt, they produce a large amount of energy that is then converted to heat. The heat boils water, creating steam that is used to turn turbines, which spins the shaft of a generator. Inside the generator, coils of wire spin in a magnetic field and electricity is produced. The Nuclear power plants use two types of reactors to achieve this process boiling water reactors (BWR) and pressurized water reactors (PWR).

The pressurized water reactor belongs to the light water type the moderator and coolant are both light water (H_2O). The cooling water circulates in two loops, which are fully separated from one another. Pressurized Water Reactors (PWR) keep water under pressure, so the water heats but does not boil even at the high operating temperature (Figure 3.8).

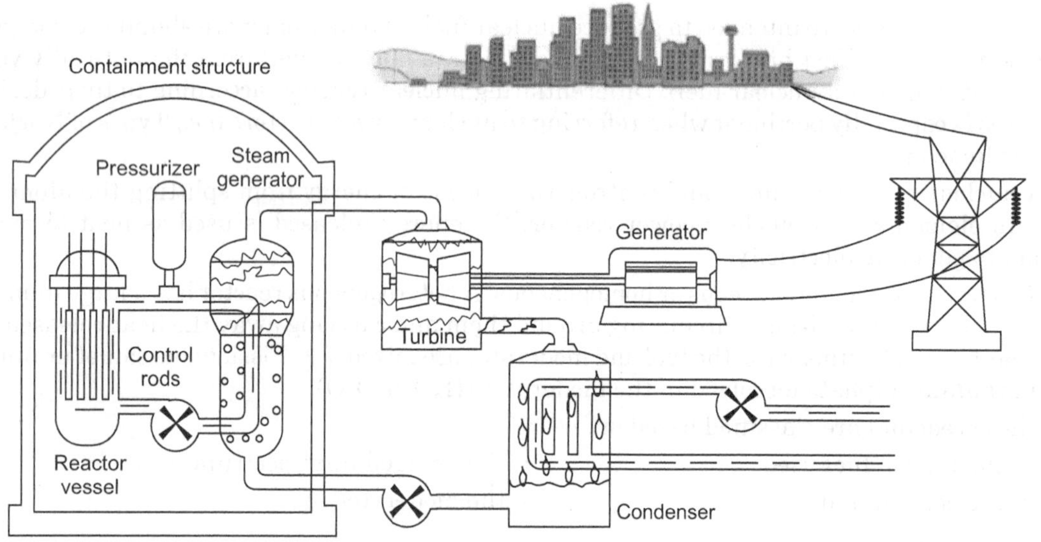

Fig. 3.8. The Pressurized Water Reactor

Constant pressure is ensured with the aid of the pressurizer. If pressure falls in the primary circuit, water in the pressurizers is heated up by electric heaters, thus raising the pressure. If pressure increases, colder cooling water is injected to the pressurizer. Since the upper part is steam, pressure will drop. The primary circuit water transferred its heat to the secondary circuit water in the small tubes of the steam generator; it cools down and returns to the reactor vessel at a lower temperature.

Since the secondary circuit pressure is much lower than that of the primary circuit, the secondary circuit water in the steam generator starts to boil. The steam goes from here to the turbine, which has high and low pressure stages. When steam leaves the turbine, it becomes liquid again in the condenser, from where it is pumped back to the steam generator after pre-heating.

Normally, primary and secondary circuit waters cannot mix. In this way, it can be achieved that any potentially radioactive material that gets into the primary water should stay in the primary loop and cannot get into the turbine and condenser. This is a barrier to prevent radioactive contamination from getting out.

In pressurized water reactors the fuel is usually low (3 to 4 percent) enriched uranium oxide, sometimes uranium and plutonium oxide mixture (MOX). In today's PWRs the primary pressure is usually 12 to 16 bars, while the outlet temperature of coolant is 30 to 32°C. PWR is the most widespread reactor type in the world they give about 64% of the total power of the presently operating nuclear power plants.

Boiling Water Reactor (BWR)

Boiling Water Reactors (BWR) heat water by generating heat from fission in the reactor vessel to boil water and create steam, which turns the generator. In both types of plants, the steam is turned back into water and can be used again in the process.

In a boiling water reactor, light water (H_2O) plays the role of moderator and coolant as well. Part of the water boils away in the reactor pressure vessel, thus a mixture of water and steam leaves the reactor core (Figure 3.9).

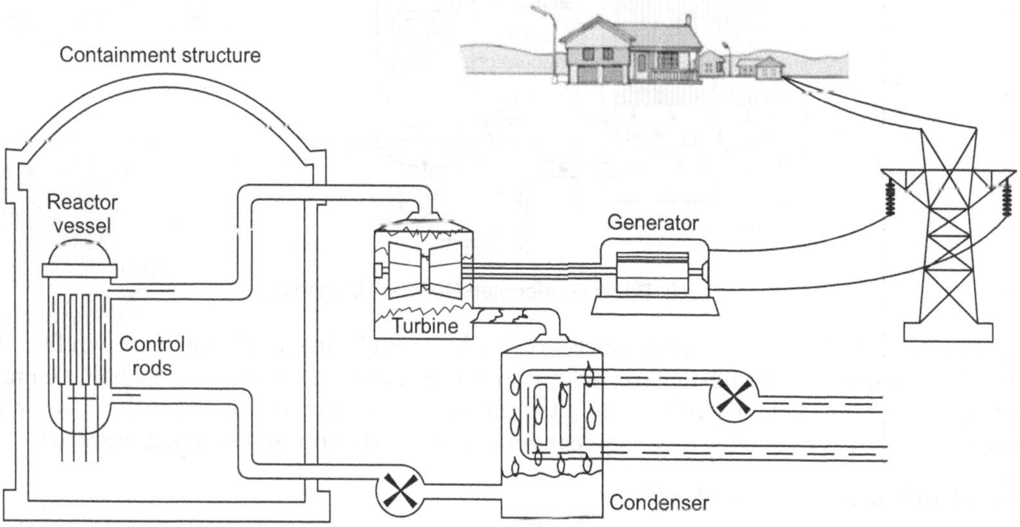

Fig. 3.9. The Boiling Water Reactor (BWR)

The so generated steam directly goes to the turbine, therefore steam and moisture must be separated. Steam leaving the turbine is condensed in the condenser and then fed back to the reactor after preheating.

Water that has not evaporated in the reactor vessel accumulates at the bottom of the vessel and mixes with the pumped back feed water.

Since boiling in the reactor is allowed, the pressure is lower than that of the PWRs. It is about 6 to 7 bars. The fuel is usually uranium dioxide. Enrichment of the fresh fuel is normally somewhat lower than that in a PWR.

The advantage of this type is that - since this type has the simplest construction - the building costs are comparatively low. 22.5% of the total power of presently operating nuclear power plants is given by BWRs.

Gas-cooled, Graphite Moderated Reactor

It belongs to commercial reactor type:

(*i*) Basic Gas-Cooled Reactor (MAGNOX)

(*ii*) Gas-Cooled Reactor (or) AGR

(*i*) Basic Gas-Cooled Reactor (MAGNOX)

Magnoxractors were built in the UK from 1956-1971 .The Magnox reactor is named after the magnesium metal. Fuel elements consisting of fuel rods encased in Magnox cans are loaded into vertical channels in a core constructed of graphite blocks (Figure 3.10).

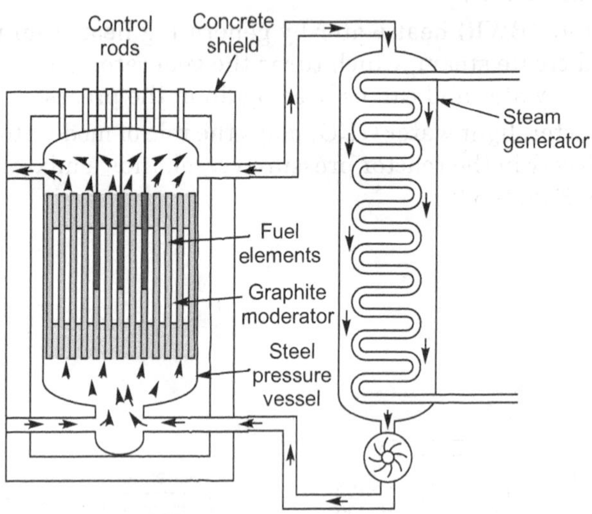

Fig. 3.10. Basic Gas-cooled Reactor (Magnox)

Further vertical channels contain control rods which can be inserted or withdrawn from the core to adjust the rate of the fission process and therefore, the heat output. The whole assembly is cooled by blowing carbon dioxide gas past the fuel cans, which are specially designed to enhance heat transfer. The hot gas then converts water to steam in a steam generator.

(*ii*) Gas-Cooled Reactor (or) AGR

Advanced Gas-Cooled Reactor (AGR) is the second generation of gas-cooled (Figure 3.11) reactors, using graphite moderator and carbon dioxide as coolant. The fuel is uranium oxide pellets, enriched to 2.5–3.5%, in stainless steel tubes.

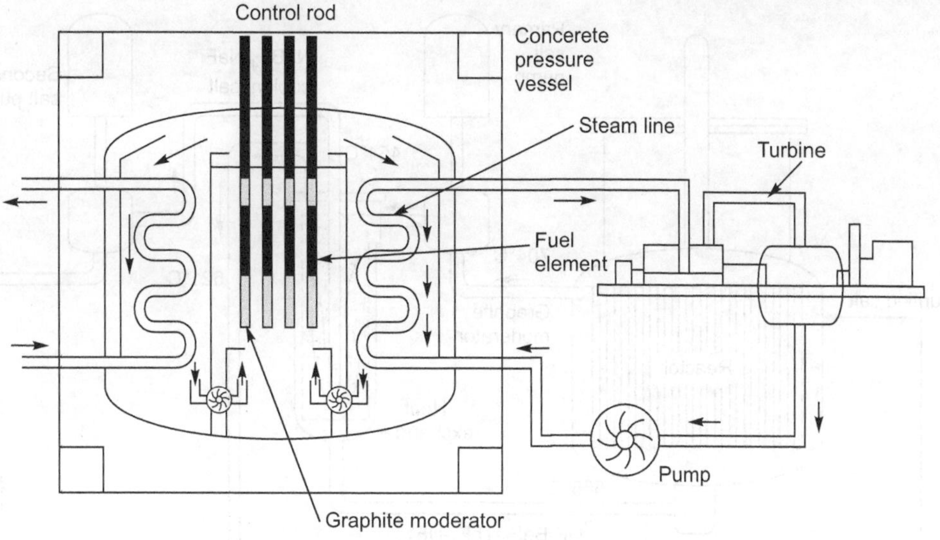

Fig. 3.11. Advanced Gas-cooled Reactor (AGR)

The carbon dioxide circulates through the core, reaching 6500°C and then past steam generator tubes outside it, but still inside the concrete and steel pressure vessel. Control rods penetrate the moderator and a secondary shutdown system involves injecting nitrogen to the coolant. They use natural uranium fuel in metal form.

Molten Salt Reactor (MSR)

A molten salt reactor (MSR) is a type of nuclear reactor where the primary coolant is a molten salt. Molten salt refers to a salt that is in the liquid phase that is normally a solid at standard temperature and pressure (STP). A salt that is normally a liquid at STP is usually called a room temperature ionic liquid, although technically molten salts are a class of ionic liquids.

Molten Salt Reactors (MSRs) are liquid-fueled reactors that can be used for (Figure 3.12) production of electricity, actinide burning, production of hydrogen, and production of fissile fuels. Electricity production and waste burn down are envisioned as the primary missions for the MSR. Fissile, fertile, and fission isotopes are dissolved in a high-temperature molten fluoride salt with a very high boiling point (140°C) that is both the reactor fuel and the coolant.

The near-atmospheric-pressure molten fuel salt flows through the reactor core. The traditional MSR designs have a graphite core that results in a thermal to epithermal neutron spectrum.

In the core, fission occurs within the flowing fuel salt that is heated to ~ 700°C, which then flows into a primary heat exchanger where the heat is transferred to a secondary molten salt coolant. The fuel salt then flows back to the reactor core.

The clean salt in the secondary heart transport system transfers the heat from the primary heat exchanger to a high-temperature Brayton cycle that converts the heat to electricity. The Brayton cycle may use either nitrogen or helium as a working gas.

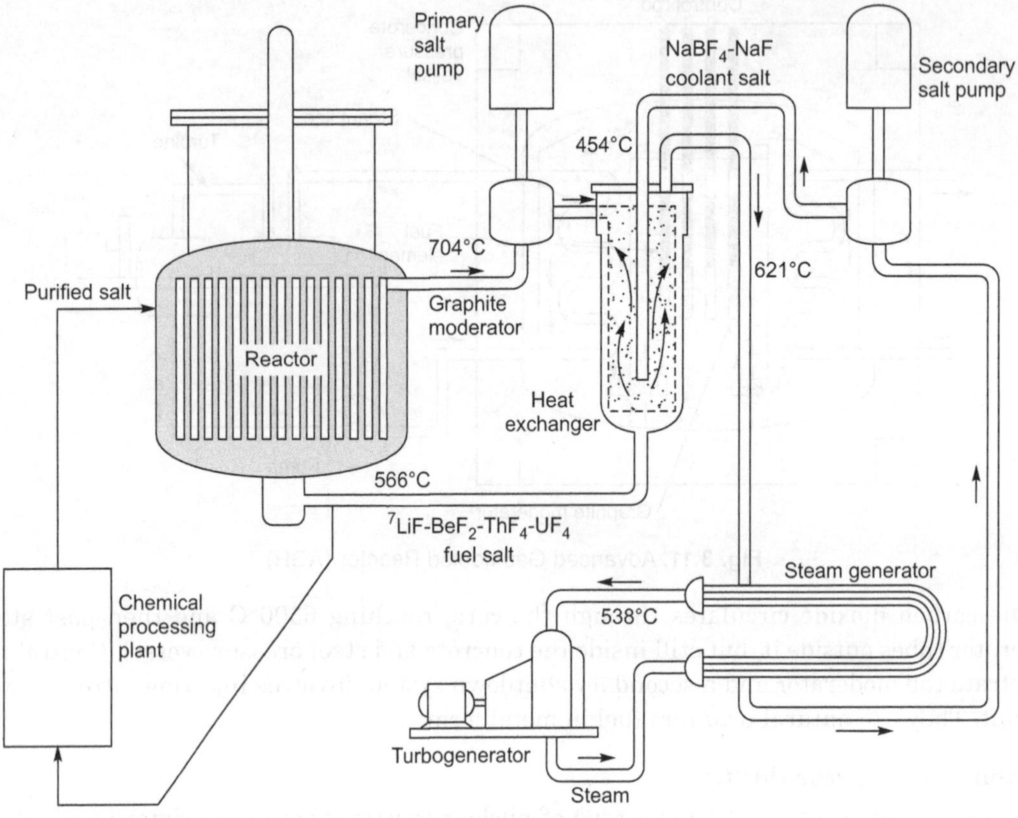

Fig. 3.12. Molten Salt Reactor

Liquid Metal Fast Breeder Reactor (LMFBR)

Definition: A type of reactor which breeds itself as well as uses liquid metal for transformation of heat in to the heat exchanger is called LMFBR.

Explanation: U-235 being less abundant cannot be sued for ever as a fuel. So a way out was needed to use the abundant isotope as a fuel in the nuclear reactor. This would increases the life span of the present Uranium reserves by 100. The sectional view of LMFBR is shown in Figure 3.13.

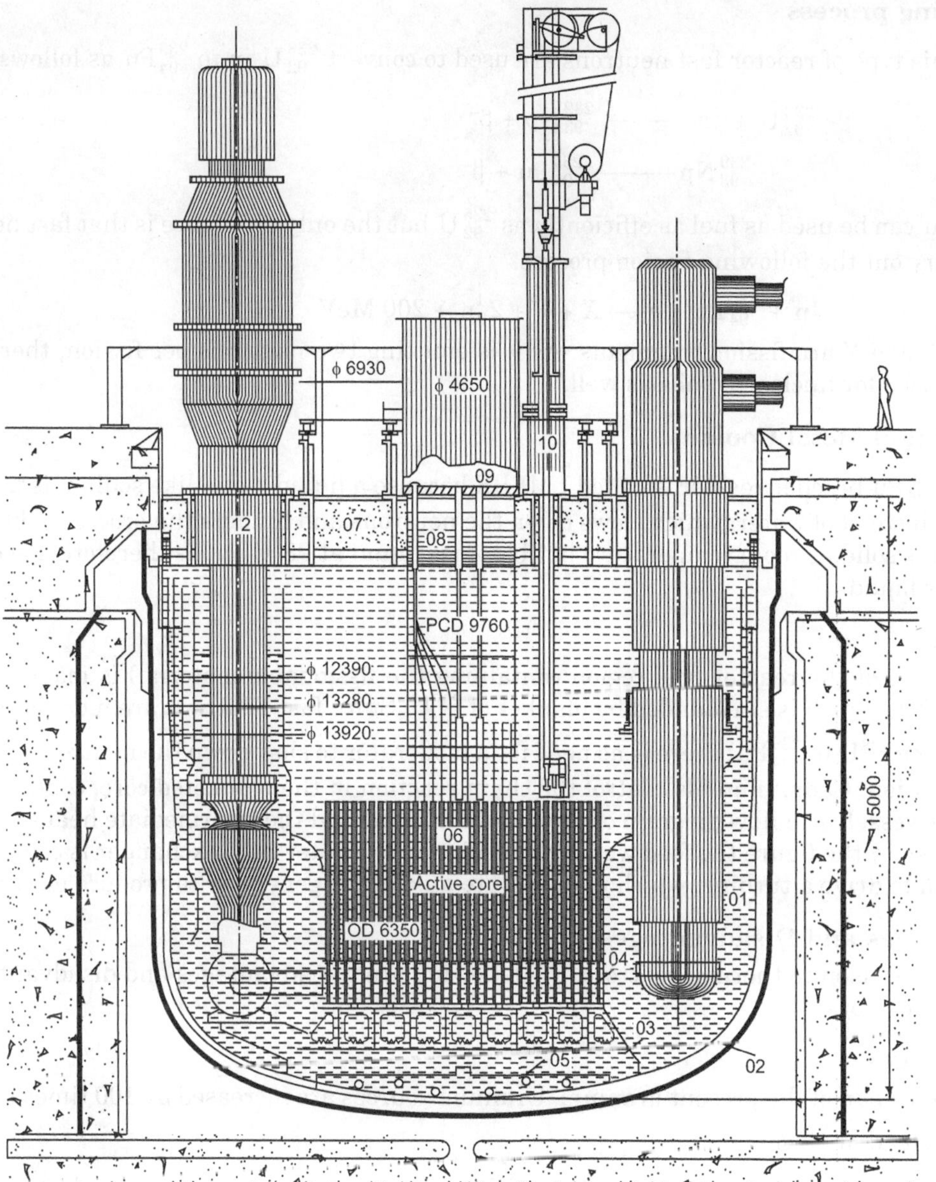

LEGEND

01. Main vessel
02. Safety vessel
03. Core support structure
04. Grid plate
05. Core catcher
06. Core
07. Top shield
08. Control plug
09. Control & safety rod drive machanism
10. In-Vessel transfer machine
11. Intermediate heat exchanger
12. Primary pump & drive

Fig. 3.13. Liquid Metal Fast Breeder Reactor

Breeding process

In this type of reactor fast neutrons are used to convert $^{238}_{92}U$ in to $^{239}_{94}Pu$ as follows:

$$^{238}_{92}U + ^{1}_{0}n \longrightarrow ^{239}_{93}Np + \beta^-$$

$$^{239}_{93}Np \longrightarrow ^{239}_{94}Pu + \beta^-$$

$^{239}_{94}Pu$ can be used as fuel as efficiently as $^{235}_{92}U$ but the only difference is that fast neutrons can carry out the following fission process.

$$^{1}_{0}n + ^{239}_{94}Pu \longrightarrow X + Y + 2^{1}_{0}n + 200 \text{ MeV}$$

where X and Y are fission fragments. As it is emitting two neutrons per fission, therefore it can be used for nuclear bombs as well.

The Liquid Metal Coolant

The $^{239}_{94}Pu$ breeder reactor is called LMFBR because a liquid metal like sodium is used as a coolant instead of water, which slows down the neutrons and hinders the process of breeding. Sodium is solid at room temperature but becomes liquid at 98°C and is therefore a good heat transfer liquid.

Working

The core of the reactor (the region containing the fissionable material) is made up of 15 to 30% ^{235}U which is surrounded by blanket of ^{238}U. Since fast neutrons are more efficient in converting ^{238}U to $^{239}_{94}Pu$, there is no need for a moderator to slow down the neutrons produced in the fission. Cadmium rods control the chain reaction in the blanketed core, contained in a primary vessel. The flow diagram of the sodium coolant and the intermediate heat exchanger are shown in the figure the feed water circulating in the outer heat exchanger is converted in to steam to drive a turbine, which may operate an electric generator to produce electricity.

Advantages and Disadvantages

In comparison to thermal reactors, the following are the advantages and disadvantages of breeder reactors:

Advantages

The life span of the present uranium. Uranium sources are increased by 100 times by using $^{238}_{92}U$ instead of $^{235}_{92}U$.

The $^{238}_{92}U$ left over while enrichment of $^{235}_{92}U$ for thermal reactor can be used.

Ores with a very small concentration of $^{235}_{92}U$ which is useless for conventional reactors can be used for the as blanket of $^{238}_{92}U$ in LMFBR.

Energy produced is of low cost due to readily availability of $^{238}_{92}U$ (without enrichment).

Disadvantages

Sodium explodes if exposed to air or water. Therefore any leakage can be dangerous. The coolant used in LMFBR and the heat exchanger is opaque. Therefore, it is not possible to see inside the core, what is going on there while repairing during maintenance.

The high energy neutron makes sodium radioactive because:

$$ {}_{0}^{1}n + {}_{11}^{23}Na \longrightarrow {}_{11}^{24}Na + \gamma\text{-ray } (T_{1/2} = 15 \text{ hrs}) $$

To be safe, it is advised to wait at least twenty halflifes which is 300 hours or 12.5 days. So it is important to wait 12.5 days after shutting down the reactor in order to carry some repair work.Long switching off the reactor, increases the cost significantly and also reliability of LMFBR.

Highly riched sodium coolant can cause a serious accident in LMFBR and then there is a chance that ${}_{94}^{239}Pu$ would be emitted in to atmosphere which can result in devastating results.

3.11 TYPES OF NUCLEAR REACTIONS

The change in the mass of particles during nuclear reactions represents the release or absorption energy. If the total mass after the reaction is reduced, the reaction releases the energy and an increase mass causes the absorption of energy.

The nuclear reactions which replace are four in nature:

1. **Inelastic Scattering:** When a neutron undergoes inelastic scattering, it is first captured by the target to form a compound nucleus, then a neutron of lower kinetic energy is expelled from the target nucleus leaving the nucleus in excited state. Thus in inelastic scattering process, part of kinetic energy of the neutron is converted into internal or excitation energy of the target nucleus. The excitation energy is subsequently emitted in the form of y-radiations.
2. **Elastic Scattering:** The situation of elastic scattering is different from inelastic scattering. In elastic scattering, the kinetic energy of the neutron is conserved. When the neutron strikes the target nucleus, imparts the part of kinetic energy to the target nucleus and its original kinetic energy is reduced. The process of elastic scattering is as "billiard ball type collision". In each collision with stationary nucleus, the neutron transfers part of its kinetic energy to the nucleus and its own kinetic 'energy is reduced. This process of scattering slows down the neutrons. The fraction of energy transferred to the target nucleus per collision depends upon the angle through which neutron is scattered and the mass of the target nucleus.

   When the neutron strikes a light nucleus such as H_2, the kinetic energy of the neutron is very much reduced and it is transferred to proton. Here the most of the neutron energy is transferred to proton as both particles have nearly same mass. With single collision of neutron with hydrogen atom nearly 75% of the neutron energy is transferred to proton. The energy transfer from the neutron to heavier target nucleus pet collision is considerably less. When a neutron collides with a carbon nucleus, the kinetic energy transferred to carbon nucleus per collision is hardly 15% of the initial kinetic energy of the neutron. The elastic scattering process plays very important role in the operation of nuclear reactors.
3. **Neutron Capture:** In this process, the colliding neutron is absorbed by the target nucleus and increases its mass number by unity. The target nucleus becomes excited and emits the energy in the form of *y*-radiations. The artificial radioactive elements are produced by this method. This type of nuclear reaction is not at all important in nuclear-power engineering.
4. **Fission:** In this type of reaction, the target nucleus absorbs thermalized (slow) neutron and becomes highly excited. Therefore, it splits into two different masses. The product

masses will also be in excited state and they try to be stable by emitting neutrons. Such type of reaction is possible only with the heavy nucleus such as ${}^{233}_{92}U$, ${}^{235}_{93}U$ and ${}^{239}_{94}Pu$.

This reaction is known as fission reaction. The nuclei produced after reaction is lighter than original nuclei and having more binding energies. The release of energy is due to an increase of mass defect of the lighter nuclei. This reaction is most important from the nuclear-power engineering.

The self-sustaining nuclear fission reaction associated with release of energy is very important in the power development.

3.12 OUTSIDE A NUCLEAR POWER PLANT

A concrete liner typically houses the reactor's pressure vessel and acts as a radiation shield. That liner, in turn, is housed within a much larger steel containment vessel. This vessel contains the reactor core, as well as the equipment plant workers use to refuel and maintain the reactor. The steel containment vessel serves as a barrier to prevent age of any radioactive gases or fluids from the plant.

An outer concrete building serves as the final outer layer, protecting the steel containment vessel. This concrete structure is strong enough to survive the kind of massive damage that might result from earthquakes or a crashing jet airliner.

These secondary containment structures are necessary to prevent the escape of radiation/ radioactive steam in the event of an accident. Workers in the control room at the nuclear power plant can monitor the nuclear reactor and take action if something goes wrong. Nuclear facilities also typically feature security perimeters and added personnel to help protect sensitive materials.

3.13 THE SIGNIFICANT PROBLEMS WITH NUCLEAR POWER PLANTS

Mining and purifying uranium has not, historically, been a very clean process. Improperly functioning nuclear power plants can create big problems. The Chernobyl disaster is a good recent example. Chernobyl was poorly designed and improperly operated, but it dramatically shows the worst-case scenario. Chernobyl scattered tons of radioactive dust into the atmosphere.

Spent fuel from nuclear power plants is toxic for centuries, and as yet, there is no safe, permanent storage facility for it.

Transporting nuclear fuel to and from plants poses some risk.

3.14 NUCLEAR POWER PLANTS - SAFETY

Safety is taken very seriously by those working in nuclear power plants. The main safety concern is the emission of uncontrolled radiation into the environment which could cause harm to humans both at the reactor site and off-site.

Radiation doses are controlled via the following procedures,

- The handling of equipment via remote in the core of the reactor
- Physical shielding
- Limit on the time a worker spends in areas with significant radiation levels
- Monitoring of individual doses and of the work environment
- Safety Mechanisms of a Nuclear Power Reactor

By regulation, the design of the nuclear reactor must include provisions for human (operator) error and equipment failure. Nuclear Plants in the western world use a Defense in Depth concept which is a system with multiple safety components, each with back-up and design to accommodate human error.

The components include:

1. Control of Radioactivity
2. Maintenance of Core Cooling
3. Maintenance of barriers that prevent the release of radiation

1. Control of Radioactivity

This requires being able to control the neutron flux. The most common way to reduce the neutron flux is include neutron-absorbing control rods. These control rods can be partially inserted into the reactor core to reduce the reactions. In modern nuclear power plants, the insertion of all the control rods into the reactor core occurs in a few second thus halting the nuclear reaction as rapidly as possible.

2. Maintenance of Core Cooling

In any nuclear reactor some sort of cooling is necessary. Generally nuclear read use water as a coolant. However some reactors which cannot use water use sodium or sodium salts.

3. Maintenance of barriers that prevent the release of radiation

There is a series of physical barriers between the radioactive core and environment. Workers are shielded from radiation via interior concrete walls. A vacuum building is connected to the reactor buildings by a pressure relief duct. The vacuum building is a 71 m high concrete structure and is kept at negative atmospheric pressure. This means that if any radiation were to leak from the reactor it would be sucked into the vacuum building and therefore prevented from being released into the environment.

3.15 WORKING PRINCIPLE OF NUCLEAR POWER PLANT

The layout of the nuclear power plant is shown in Figure 3.14.

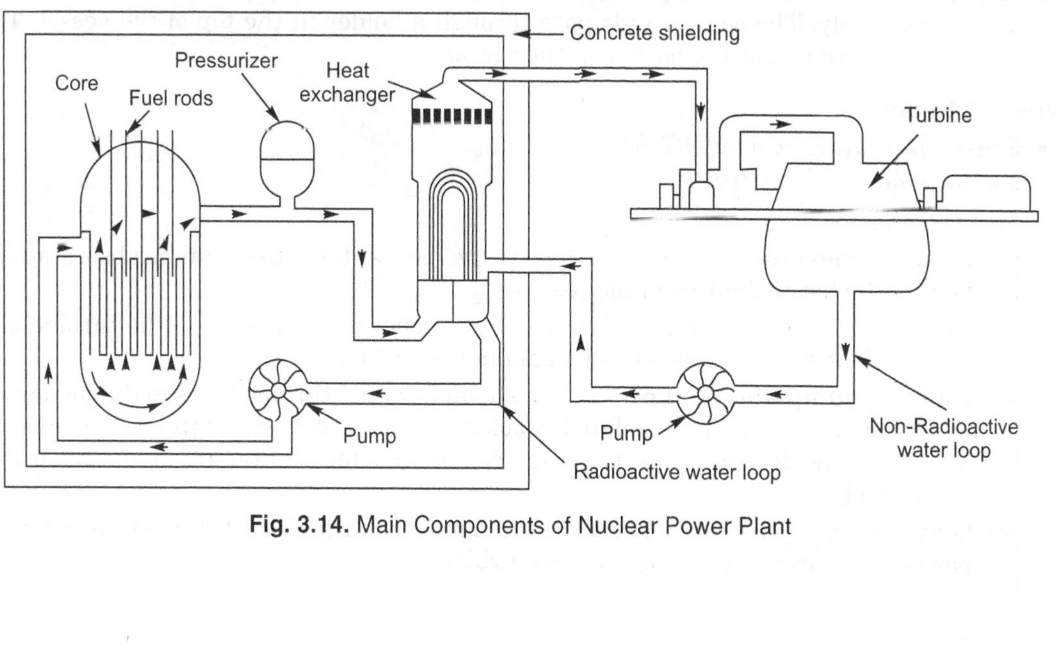

Fig. 3.14. Main Components of Nuclear Power Plant

The main components are:

(*a*) **Nuclear reactor:** It has the following components:

1. Core (fuel rods)
2. Moderator
3. Control rods
4. Coolant
5. Reflector
6. Radiation shield
7. Reactor vessel

1. **Core:** It consists of the nuclear fuel and space for the coolant. The nuclear fuels generally used are U-233, U-235 and Pu-239. The fuel is shaped and located such that the heat released in the reactor is uniform.
2. **Moderator:** It is used to reduce the speed of the fast moving neutrons. Graphite, heavy water or beryllium is used as moderators for the natural uranium. Ordinary water is used as moderator only with the enriched uranium.
3. **Control rods:** They are used to maintain the chain reaction at the required level and to start the chain reaction when starting cold and to shut down in case of emergency. The control rods are made of cadmium, boron and hafnium.
4. **Coolant:** Coolant transfers the heat produced in the reactor to the steam generator for raising the steam. The coolant employed are ordinary and heavy water, air, carbon dioxide, helium and hydrogen and liquid metals like sodium, potassium, etc.
5. **Reflector:** It is used to reflect the escaping neutrons back into the core. It improves the neutron economy of the reactor. The reflectors are heavy water, graphite and beryllium.
6. **Radiation shield:** It is a concrete shield used to absorb the hazardous radiations like alpha, beta, gamma rays and neutrons which tend to come out of the reactor to the atmosphere.
7. **Reactor vessel:** It is housing for all the components and is designed to withstand high pressures safely. The control rods pass through a holder in the top of the vessel. The reactor is positioned at the bottom of the vessel.

Types of Reactors

- Pressurized water reactor (PWR)
- Boiling water reactor (BWR)
- Heavy water cooled reactor

(*b*) **Steam generator:** It is used to convert the feed water into steam by absorbing the heat of the hot coolant from the reactor.

(*c*) **Turbine:** The steam produced in the steam generator is passed to the turbine and work is done by the expansion of steam in the turbine.

(*d*) **Coolant pump and feed pump:** The steam from the turbine flows to the condenser where cooling water is circulated. Coolant pump and feed pump are provided to maintain the flow of coolant to the reactor and feed water to steam generator respectively.

(*e*) **Generator:** It is used to convert the mechanical energy of the turbine into electrical power. It is directly coupled with the turbine.

Advantages

1. A nuclear power plant occupies less space when compared with other conventional power plants of same size.
2. Fuel transportation cost and fuel storage facilities needed are less.
3. They are not affected by adverse weather conditions.

Disadvantages

1. Cost of establishing nuclear plant is more than that for a hydro thermal power plant.
2. Sufficient care must be taken to dispose off the radio active wastes which may otherwise pose a serious problem to the health of the workers as well as to the environment.
3. Maintenance cost is high.
4. It requires skilled personnel for operation.

3.16 GENERAL COMPONENTS OF NUCLEAR REACTOR

The nuclear reactor may be regarded as a substitute for the boiler fire box of steam plant or combustion chamber of a gas-turbine plant. The heat produced in the nuclear power plant is by fission whereas in steam and gas turbine plants, the heat is produced by combustion. The other cycle of operation and components required are exactly same either as steam plant if steam is generated by using the heat of fission or a gas turbine plant (closed or open type) if gas is heated by using the heat of fission. The steam or the gas may be used as working fluid in nuclear power plant. The nuclear power plant may be of steam driven turbine or gas driven turbine as per the choice of the fluid.

The general arrangement of Nuclear power plant with essential components using steam as working fluid is shown in Figure 3.15.

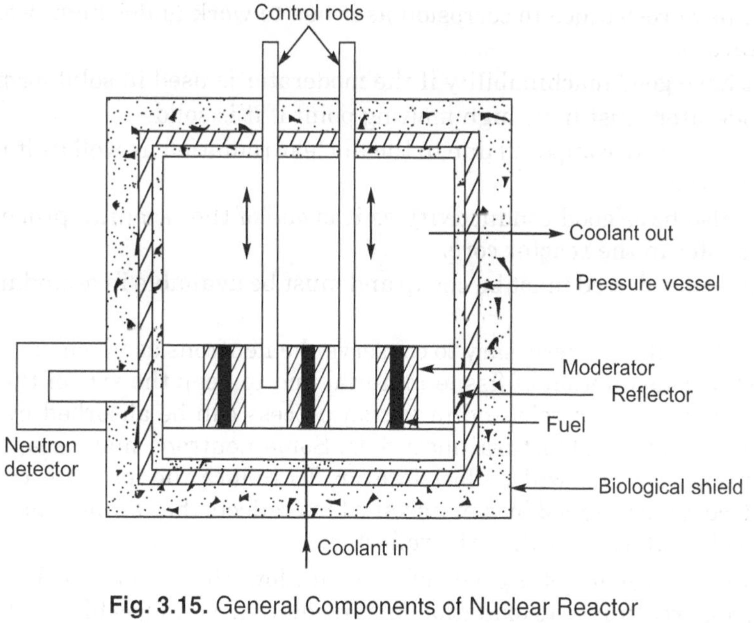

Fig. 3.15. General Components of Nuclear Reactor

1. **Fuel.** The nuclear fuels which are generally used in reactors are ${}^{235}_{92}U$, ${}^{239}_{94}Pu$ and ${}^{233}_{92}U$. Among the three, the ${}^{235}_{92}U$ is naturally available upto 0.7% in the uranium ore and the

remaining is ${}^{238}_{n}U$. The other two fuels ${}^{239}_{94}Pu$ and ${}^{233}_{92}U$ are formed in the nuclear reactors during fission process from ${}^{238}_{n}U$ and ${}^{232}_{90}Th$ due to the absorption of neutron without fission. The fuel is shaped and located in the reactor in such a manner that the heat production within the reactor is uniform. The fuel elements are designed taking into account the heat transfer, corrosion and structural strength.

In homogeneous reactors, the fuel and moderator are mixed to form a uniform mixture *i.e.* uranium and carbon and then it is used in the form of rods or plates in the reactor core. In heterogeneous reactors, the fuel is used in the form of rods or plates and moderator surrounds the fuel elements. This arrangement is commonly used in most of the reactors. The fuel rods are clad with aluminum, stainless steel or zirconium to prevent the oxidation of uranium.

2. **Moderator.** The moderator is a material which reduces the kinetic energy of fast neutron (1 MeV or 13200 km/sec) to slow neutron (0.25 eV or 2200 m/sec) and this is done in a fraction of second. The fission chain-reaction in the nuclear reactor is maintained due to slow neutrons when the ordinary uranium is used as fuel. The function of the moderator is to increase the probability of reaction. The slowing down of neutrons is effectively done by the light elements as H_2, D_2, N_2, O_2, C and Be as mentioned earlier. The graphite, heavy water or beryllium can be used as moderator with natural uranium. The ordinary water is used as moderator only with the enriched uranium for the reason mentioned earlier.

   The desirable properties of a good-moderator are listed below:

   1. It must be as light as possible, as slowing down action is more effective in elastic collision with light elements.
   2. The moderator must be able to slow down the neutrons earliest possible but it must not absorb them (low absorption cross-section).
   3. It must have resistance to corrosion as it has to work under high pressure and high temperature.
   4. It must have good machinability if the moderator is used in solid form.
   5. The moderator must have high melting point if it is solid.
   6. It should not be decomposed due to the nuclear radiation as well as it must have high chemical stability.
   7. It must also have good conductivity as it is one of the essential properties for better heat transfer in the reactor core.
   8. The material selected must be cheap and must be available in abundance and in pure from.

3. **Reflector**: It is always necessary to conserve the neutrons as much as possible in order to reduce the consumption of fissile material and to keep the size of the reactor small. The neutrons which are released in fission process can be absorbed by the fuel itself, moderator, coolant or structural materials. Some neutrons may escape from the core without absorption and will be lost forever. To reduce the loss of escape, the reactor inner surface is surrounded by a material which reflects the escaping neutrons back into the core. This material is called the reflector.

   The required properties of a good reflector are low absorption and high reflection for neutrons, high resistance to oxidation and irradiation as well as high radiation stability.

   Many times the material used as moderator is also used as reflector because the moderating materials have good reflecting characteristics. The H_2O. D_2O and carbon are also used as reflectors. The amount of fissionable material required can be reduced with the use of good reflector.

It is necessary to provide some method of cooling the reflector as it gets heated due to collision of neutrons with its atoms.

4. **Coolant.** The main purpose of the coolant in the reactor is to transfer the heat produced inside the reactor. The same heat carried by the coolant is used in the heat exchanger for further utilization in the power generation.

   Some of the desirable properties of a good coolant are listed below:

   1. It must not absorb the neutrons.
   2. It must have high chemical and radiation stability.
   3. It must be non-corrosive.
   4. It must have high boiling point (if liquid) and low melting point (if solid).
   5. It must be non-oxidising and non-toxic.
      The above-mentioned properties are essential to keep the reactor core in safe condition as well as for the better functioning of the coolant.
   6. It must also have high density, low viscosity, high conductivity and high specific heat. These properties are essential for better heat transfer and low pumping power.

   The water, heavy water, gas (He, CO_2), a metal in liquid form (Na) and an organic liquid are used as coolants.

   The coolant not only carries large amounts of heat from the core but also keeps the fuel assemblies at a safe temperature to avoid their melting and destruction.

5. **Control Rods.** Some type of control is absolutely necessary to fulfil the following functions:

   (*a*) To start the nuclear chain reaction when the reactor is started from cold.

   (*b*) The chain reaction should be maintained at steady state condition (controlled chain reaction) at the required level.

   (*c*) To shut down the reactor automatically under emergency condition, *i.e.*, a pump circulating the coolant through the reactor fails.

   The control is necessary to prevent the melting of fuel rods, disintegration of coolant and destruction of reactor as the amount of energy released is enormous.

   The control of the chain reaction is possible either by removing the fuel rods or the neutrons continuing the chain reaction from the core of the reactor. It is always easy to get rid of the neutrons. It is generally done by inserting the control rods into the fuel tubes as shown in Figure 3.15.

   The materials used for control rods must have very high absorption capacity for neutrons. The common materials used for control rods are cadmium, boron or hafnium. The action of control rods can be very well compared with the action of the blotting paper which absorbs extra ink without spreading.

6. **Shielding.** The reactor is a source of intense radioactivity as mentioned earlier and these radiations are very harmful to the human life. The common radiations from the reactor are α-rays, β-rays, γ-rays and fast neutrons. To prevent the effects of these radiations on the human life, it is necessary to absorb them before emitting to the atmosphere.

   Neutrons, γ-rays and all other radiations are effectively absorbed by the concrete and steel. The inner lining of the core is made of 50 to 60 cm thick steel plate and it is further thickened by few metres using concrete. The lining of steel plate absorbs these energies and becomes heated but prevents the adjacent wall of reactor vessel from becoming heated. The therm 21 shield (steel plate) is cooled by the circulation of water.

7. **Reactor vessel.** The reactor vessel encloses the reactor core, reflector and shield. It also provides the entrance and exit passages for directing the flow of coolant. The reactor vessel has to withstand the pressure as high as 200 bar or above. The holes at the top of the vessel are provided to insert the control rods. The reactor core (fuel and moderator assembly) is generally placed at the bottom of the vessel.

3.17 MERITS AND DEMERITS OF NUCLEAR POWER PLANT

Merits:

- Very small quantity of fuel is required for power generation
- Requires less space
- Low operating cost
- The cost of nuclear power will not be affected by fluctuations in petrol, diesel and gas prices since the nuclear power plant does not use fossil fuel
- Nuclear fuels are available in plenty around the world
- Reliability of operation
- Does not require large quantities of water for cooling
- Economical for producing electricity in bulk
- Environment is affected to a minimum extent only.

Demerits

- Fuel is expensive and hard to recover
- Capital cost is very high
- Erection and commissioning of plant requires high technical knowledge
- Nuclear radiation can be extremely dangerous
- Fission by-products are generally radioactive and may cause nuclear pollution. Maintenance cost is high
- Not suited for varying loads as reactor does not respond to fluctuations. Disposal of nuclear by-product is difficult and problematic. The by-product emits radiation as well as heat and can turn to be catastrophic
- Spent nuclear fuel is radioactive and extremely dangerous, requiring a substantial infrastructure to secure the power plant and to handle the fuel
- Transporting nuclear fuel posses a serious contamination risk.

3.18 NUCLEAR POWER PLANTS IN INDIA

1. Tarapur Nuclear Power Station (Near Bombay)
2. Rana Prathap Sagar Nuclear Power Station (Rajasthan)
3. Kalpakkam Nuclear Power Station (Tamil Nadu)
4. Navora Nuclear Power Station (Uttar pradesh)
5. Kakrapar Nuclear Power Station (Gujarat)
6. Kaiga Nuclear Power Station (Karnataka)

REVIEW QUESTIONS

Two Marks Questions

1. What is the principle of nuclear energy?
2. What are the benefits of nuclear energy?
3. Define Elements and Atoms.
4. What are the main components of nuclear power plant?
5. What are the types of reactors?
6. What are the advantages and disadvantages of nuclear power plant?
7. What is mean by radioactivity?
8. What is radioactive decay?
9. What are characteristics of radioactive decay?
10. What are the biological effects of radiation?
11. What is fusion?
12. What are the types of nuclear reactions?
13. Define nuclear fission.
14. What are the factors must be taken into account during nuclear fission?
15. What are the conditions satisfied to sustain nuclear fission process?
16. Mention the various types of fast breeders.
17. What factors control the selection of particular type of reactor?
18. Give the range of power and fuel cycle in sodium cooled reactors.
19. Differentiate nuclear fission and fusion.
20. Define multiplication factor.
21. What is nuclear reactor?
22. Define chain reaction and give its requirements.
23. What is halflife of nuclear fuels?
24. What is the function of pressurizer in pressured water reactor?
25. What is the function of rod?
26. State Daltons atomic theory.
27. Define the term breeding.
28. What are the advantages and disadvantages of fast breeder reactor?
29. What are the advantages and disadvantages of gas cooled reactor?
30. List out nuclear waste disposal methods.
31. What are the advantages of sodium fast reactor?
32. What are the factors for selecting site for nuclear power station?
33. What are the advantages and disadvantages of CANDU reactor?
34. What are the advantages and disadvantages of BWR?
35. What is LMFBR? Why the liquid metal is preferred as coolant in a reactor?
36. What are the advantages and disadvantages of SCWR?
37. What is known as binding energy?
38. What is an isotope?
39. Describe moderator and reflector.
40. What are the desirable properties of good coolant?
41. What is meant by shielding?
42. Write two merits and demerits of nuclear power plant.
43. What are the nuclear power plants in India?

Sixteen Marks Questions

1. What is known as nuclear fission? What are the essential requirements to cause nuclear fission?
2. With a neat sketch explain the boiling water reactor power plant.
3. With a neat sketch explain the working of pressurized water reactor in nuclear plant.
4. Explain with neat sketch about the controlled and uncontrolled nuclear chain reaction.
5. Explain the working principal of nuclear power plant with neat layout.
6. Explain the construction and working of super critical water reactors.
7. Explain the working principal of general components of nuclear reactor with a suitable sketch.
8. Explain the principle of operation of fast breeder reactor and CANDU reactor.
9. Explain the working principle of LMFBR and list out the advantages and disadvantages.
10. Explain the types of nuclear reactions.
11. Explain the nuclear power plants- Safety.
12. What is radioactive decay? Explain the types of radioactive decay with suitable sketch.
13. Explain nuclear life cycle with a neat layout.
14. Explain Uranium-235 chain reaction with suitable sketch.
15. What are the types of radiation associated with nuclear fission? Explain.
16. Explain the merits and demerits of nuclear power plant.
17. Explain the following with neat sketch:
    (*i*) Gas Cooled reactor (*ii*) Molten Salt reactor

4

Gas Power Plant

4.1 GAS POWER PLANT

4.1.1 Introduction

The idea of gas turbine is the oldest one, and its working principle is an improved version of the wind mill, which was used several centuries back. In order to achieve an efficient working of the turbine, the movement of gas (or air) is properly controlled and then directed on the blades fixed to the turbine runner. The air under pressure is supplied to the turbine by an air compressor, which is run by the turbine itself.

In a gas turbine, first of all, the air is obtained from the atmosphere and compressed in an air compressor. The compressed air is then passed into the combustion chamber, where it is heated considerably. The hot air is then made to flow over the moving blades of the gas turbine, which imparts rotational motion to the runner. During this process, the air gets expanded and finally it is exhausted into the atmosphere. A major part of the power developed by the turbine is consumed for driving the compressor (which supplies compressed air to the combustion chamber). The remaining power is utilized for doing some external work.

4.1.2 Classification of Gas Turbines

Though the gas turbines may be classified in many ways, yet the following are important from the subject point of view:

1. According to path of the working substance
   (*a*) Closed cycle gas turbines, (*b*) Open cycle gas turbines, and
   (*c*) Semi-closed gas turbines
2. According to process of heat absorption
   (*a*) Constant pressure gas turbines, and
   (*b*) Constant volume gas turbines

We shall discuss all the above mentioned gas turbines one by one.

4.1.3 Closed Cycle Gas Turbine Power Plant

A closed cycle gas turbine, in its simplest form, consists of a compressor, heating chamber, gas turbine which drives the generator and compressor, and cooling chamber. It is a closed cycle power plant (Figure 4.1) in which the heat of the hot exhaust gases from the turbine is utilized to preheat the air entering the combustion chamber.

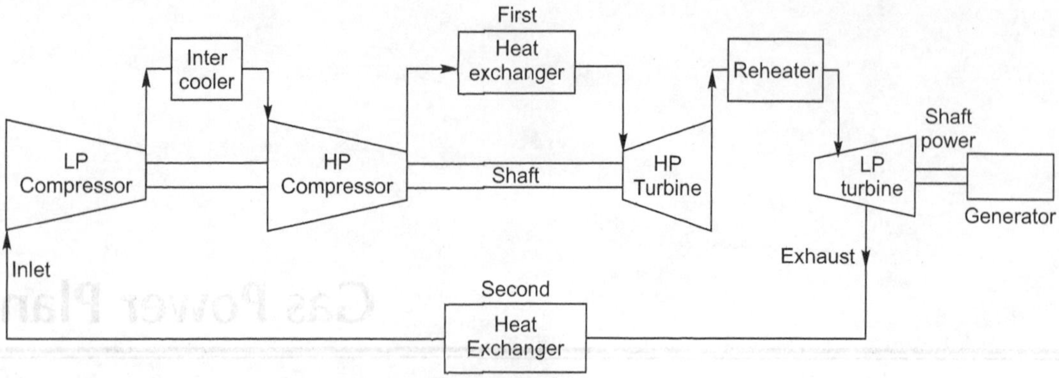

Fig. 4.1. Gas Turbine Power Plant – Closed Cycle

It consists of the following elements.

1. **Low Pressure (LP) Air Compressor:** Air is drawn from the atmosphere and passed to the LP air compressor through the air filter. Air is compressed at two stages to get the compressed air at required pressure. If it is fully compressed is one stage, then the power required to run the compressor will be around 66% of the total power produced by the turbine. To reduce the power required to run the compressor, the air is compressed in two stages (*i.e.,*) in low pressure and high pressure compressors. The air is first compressed to low pressure in LP air compressor and then this compressed air is sent to HP air compressor through the intercooler.
2. **Intercooler:** The power required to compress the air is directly proportional to the air temperature at inlet. So, if intercooling is carried out between the two stages of compression, then the total work can be reduced. Therefore, the compressed air from the LP compressor is cooled and then passed to HP compressor for further compression.
3. **High Pressure (HP) Air Compressor:** The compressed air from the intercooler is further compressed to a high pressure. Then it is passed into the regenerator.
4. **Regenerator:** In the closed cycle system, the heat of the turbine exhaust gases will be again utilized. The hot gases will not be allowed to go as waste. To make use of this heat, a regenerator is provided in the system. In the regenerator, the heat of the hot exhaust gases from the turbine is used to preheat the air entering the combustion chamber.
5. **Combustion Chamber:** Hot air from regenerator flow to the combustion chamber. Fuel of any kind like kerosene, gasoline, natural gas is injected into the combustion chamber. This fuel is burnt in the presence of hot air from the regenerator. The products of combustion at high temperature and pressure are passed to high pressure turbine.
6. **High Pressure (HP) Turbines:** The hot pressurized gases from the combustion chamber are expanded in HP turbine. On expanding the gases on the turbine blades, the rotor of the turbine will be rotated. This mechanical work will be used to run a generator to produce electrical power.
7. **Low Pressure (LP) Turbines:** The output of the HP turbine will not be sent as waste. It will be again sent to another combustion chamber where the exhaust gases from the HP turbine will be used to burn the fuel again. The resulting products of combustion will be passed to the LP turbine where they will be expanded. The heat from the exhaust gases of the HP turbine is thus utilized.

The output gases of the LP turbine also contain some heat. This heat also will not be wasted. This will be used in the regenerator to preheat the air entering the combustion chamber before the HP turbine. The reheating combustion chamber placed between the HP and LP turbines, the regenerator, the intercooler are the important elements of the gas turbine to increase the overall efficiency of the plant.

4.1.4 Open Cycle Gas Turbine Power Plant

The schematic arrangement of an open cycle gas turbine is shown in Figure 4.2. In this turbine, the air is first sucked from the atmosphere and then compressed isentropically (generally in a rotary compressor) and then passed into the combustion (*i.e.*, hot gases formed by the combustion of fuel) also get mixed up with the compressed air, this increasing the mass of compressed air. The hot gas is then made to flow over the turbine blades (generally of reaction type). The gas, while flowing over the blades, get expanded and finally exhausted into the atmosphere.

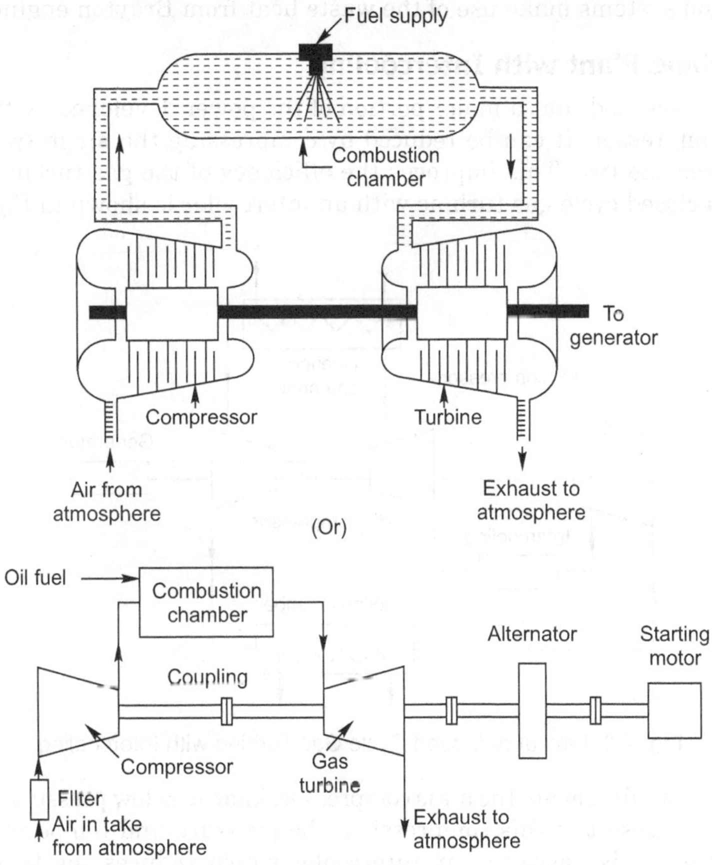

Fig. 4.2. Gas Turbine Power Plant – Open Cycle

An open cycle gas turbine is also called a continuous combustion gas turbine as the combustion of fuel takes place continuously. This turbine also works on Joule's cycle. The relation for work done by the compressor and turbine are same as those of closed cycle gas turbine.

If the fuel is burnt by atmospheric air in the combustion chamber and the gases coming out of the turbine are exhausted to the atmosphere, without fully utilizing the heat of the products of combustion, then the system is called open cycle system.

Efficiency

The Brayton cycle is used to analyze the gas turbine power plant. The thermal efficiency of a simple cycle gas turbine plant can be increased by using one of the following methods:

(1) By reducing the work required to run the compressor

(2) By reducing the heat (fuel) supplied in the combustion chamber.

The efficiency of a Brayton engine can be improved in the following manners:

1. Intercooling
2. Regeneration
3. Combined cycle system which combines with a Rankine engine to further increase overall efficiency.
4. Cogeneration systems make use of the waste heat from Brayton engines.

4.1.5 Gas Turbine Plant with Intercooling

We have already discussed that a major portion of the power developed by the gas turbine is utilized by the compressor. It can be reduced by compressing the air in two stages with an intercooler between the two. This improves the efficiency of the gas turbine. The schematic arrangement of a closed cycle gas turbine with an intercooler is shown in Figure 4.3.

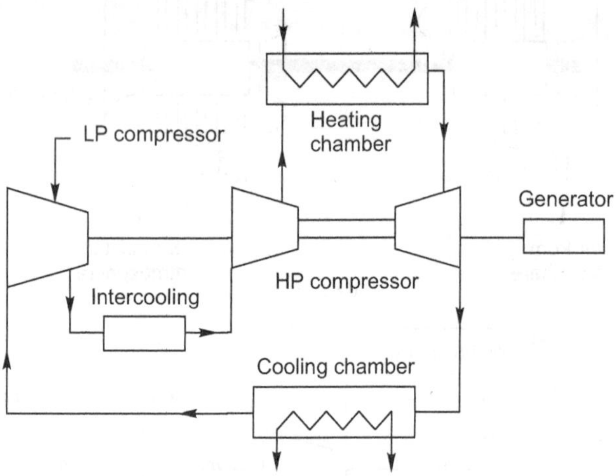

Fig. 4.3. Layout of Closed Cycle Gas Turbine with Intercooling

In this arrangement, first of all, the air is compressor, known as low pressure (LP) compressor. We known that as a result of this compression, the pressure and temperature of the air is increased. Now the air is passed to an intercooler which reduces the temperature of the compressed air to its original temperature, but keeping the pressure constant. After that, the compressed air is once again compressed in the second compressor known as high pressure (HP) compressor. Now the compressed air is passed through the heating chamber and then through the turbine. Finally, the air is cooled in the cooling chamber and again passed into the low pressure compressor as shown in Figure 4.3.

4.1.6. Gas Turbine Plant with Reheating

The output of a gas turbine can be considerably improved by expanding the hot air in two stages with a reheater between the two. The schematic arrangement of a closed cycle gas turbine with reheating is shown in Figure 4.4.

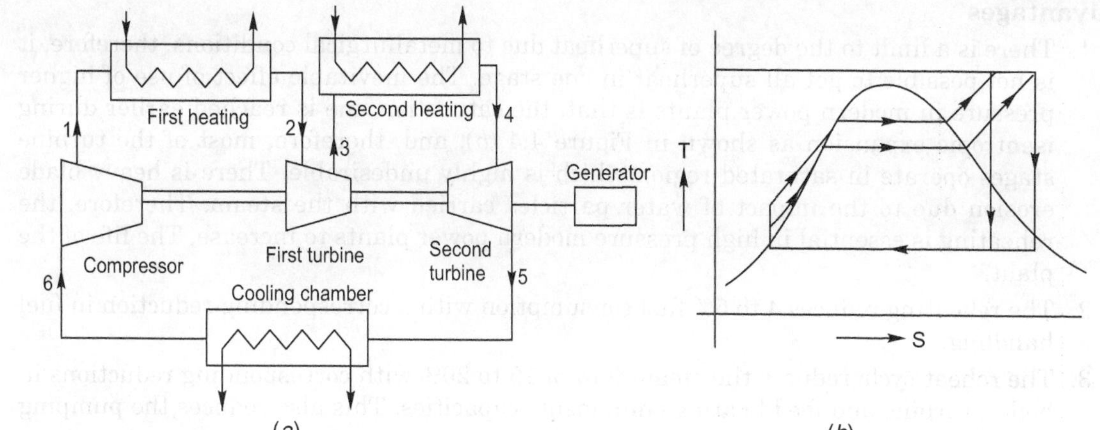

Fig. 4.4. Layout of Closed Cycle Gas Turbine with Reheating

In this arrangement, the air is first compressed in the compressor, passed into the heating chamber, and then to the first turbine. The air is once again passed on to another heating chamber and then to the turbine. Finally, the air is cooled in the cooling chamber and again passed into the compressor as shown in Figure 4.4 (*a*).

In reheat cycle, the steam is extracted from a suitable point in the turbine and is reheated with the help of the flue gases in the boiler furnace as shown in Figure 4.4. The main purpose of reheating is to increase the dryness friction of steam passing through the lower stages of the turbine. The dryness fraction of steam coming out of turbine should not fall below 0.88. By using the reheat cycle, the specific steam consumption decreases and thermal efficiency also increases. The increase in thermal efficiency due to reheat depends upon the ratio of reheat pressure to original pressure of steam. The reheat pressure is generally kept within 20% of the initial pressure of the steam. The efficiency of the reheat cycle may be less than the Rankine efficiency if the reheat is used at low pressure.

The main advantage of reheat cycle is to reduce the specific steam consumption and consequently reduce the size of the boiler and auxiliaries for the same output.

The reheat cycle is only preferred for high capacity plants (above 100 MW and when pressure of the steam is as high as 100 bar. Only one stage reheating is generally used in practice.

It is not preferred for low capacity plants as the cost of the reheater is not justified.

Referring to Figures 22.3 and 22.4, the total heat supplied and work done are given by

$$hs = (h_1 - h_{f6}) + (h_3 - h_2)$$

and

$$W = (h_1 - h_2) + (h_3 - h_4) - (h_{f6} - h_{f5})$$

where $(h_{f6} - h_{f5})$ is the pump work.

Therefore, the efficiency of the reheat cycle is given by,

$$(\eta_{\text{reheat}}) = \frac{(h_1 - h_2) + (h_3 - h_4) - (h_{f6} - h_{f5})}{(h_1 - h_{f6}) + (h_3 - h_2)}$$

Advantages and Disadvantages of Reheat Cycle

The advantages of reheat cycle over simple Rankine-cycle for turbine generator having the same rating and the same steam condition at stop-valve are listed below.

Advantages

1. There is a limit to the degree of superheat due to metallurgical conditions, therefore, it is not possible to get all superheat in one stage. The inevitable effect of use of higher pressure in modern power plants is that, the saturation line is reached earlier during isentropic expansion as shown in Figure 4.4 (*b*), and, therefore, most of the turbine stages operate in saturated region which is highly undesirable. There is heavy blade erosion due to the impact of water particles carried with the steam. Therefore, the reheating is essential in high pressure modern power plants to increase, The life of the plant.
2. The reheating reduces 4 to 5% fuel consumption with a corresponding reduction in fuel handling.
3. The reheat cycle reduces the steam flow of 15 to 20% with corresponding reductions in boiler, turbine and feed heating equipments capacities. This also reduces the pumping power in that proportion.
4. The wetness of the exhaust steam with reheat cycle is reduced to 50% of Rankine cycle with a corresponding reduction in exhaust blade erosion.
5. Lower steam pressures and temperatures and less costly materials can be used to obtain the required thermal performance.
6. A reduction in steam volume and heat to the condenser is reduced by 7 to 8%. Therefore, the condenser size and cooling water requirement are also reduced by the same percentages.
7. The size of the LP turbine blades is reduced because specific volume is reduced by 7 to 8%.

The advantages claimed for the reheat cycle are higher thermal efficiency, reduced feed pump power, smaller condenser, smaller boiler, long life of turbine and less handling of fuel and firing equipment.

Disadvantages

1. The cost of extra pipes and equipments and controls makes this cycle more expensive than ordinary Rankine cycle. Therefore, the minimum capacity of the plant must be 10 MW for the adoption of reheat cycle.
2. The greater floor space is required to accommodate the longer turbine and reheat piping.
3. The turbine blades of the second stage immediately after the reheater is considerably longer than the corresponding pressure stages of non-reheat set. Therefore, additional design problems are involved.
4. The complexity of operation and control increases with the adoption of reheat cycle.
5. At light loads, the steam passing through the last blade rows to the condenser are seriously superheated if the same reheat is maintained. Feed water is sometimes sprayed into the low pressure cylinders as low steam flows as a precaution against overheating of blades.

The above disadvantages are compensated in new installations by the saving in cost of the equipments for coal storage, coal handling, burning, fans and dust collectors, ash disposal, boiler use structure and auxiliary electrical equipments.

4.1.7 Gas Turbine Plant with Regeneration

The Rankine efficiency of a steam cycle is less than Carnot cycle as all the heat is not supplied at the highest temperature. The heating of feed water from 3 to 4 takes place at lower temperature. The Rankine cycle will be as efficient as Carnot if the temperature of feed water is raised to boiler temperature by *reversible interchange* of heat before it enters the boiler.

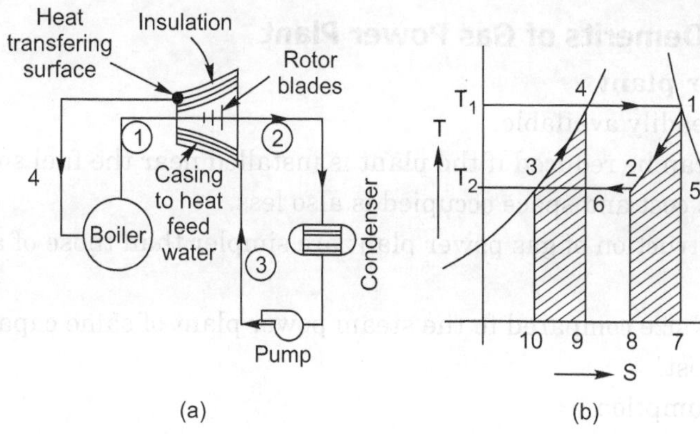

Fig. 4.5. An Arrangement for Reversible Heat Transfer to Feed Water

An arrangement for reversible heat transfer to feed water before entering into the boiler is shown in Figure 4.5 (*a*) and its corresponding *T-S* diagram is shown in Figure 4.5 (*b*). The dry-saturated steam coming out from boiler enters into the turbine at temperature T_1 and it expands adiabatically to temperature T_2. The condensate coming out from condenser is pumped back through the annular space of the turbine as shown in figure. The feed water is heated with the help of steam in a reversible manner, the temperature of steam and water is same at any section.

Such type of heating is known as *Regenerative Heating* as the steam is used to heat itself (feed water). Under ideal regeneration condition, the feed water enters at temperature T_1 and represented by the point 4. The temperature of steam entering the turbine and temperature of water leaving the turbine are same. If the system is considered as isolated adiabatic system, the heat lost by the steam must be equal to heat gained by the water.

$\therefore$ Area "3 – 4 – 9 – 10 – 3" = Area "1 – 7 – 8 – 2 – 1".

This type of heating arrangement (non-extraction feed heating) gives the efficiency equivalent to Carnot cycle efficiency. This type of arrangement cannot be used in practice because steam becomes too wet in the latter stages of the turbine. In actual practice, the advantage of regenerative heating principle is taken by "bleeding" a part of steam from the turbine at certain stages of expansion and it is used for heating the feed water in the separate feed heaters. This arrangement does not reduce the dryness fraction of remaining steam passing through the turbine.

If there were an infinite number of these extraction feed heating stages, the resulting cycle would be thermodynamically equivalent to the non-extraction feed heating cycle. The use of regenerative feed heating enabled the Rankine cycle to be improved upon and a nearer approach made to Carnot cycle. To heat the condensate from the hot-well temperature to a temperature corresponding to the saturation temperature of steam at the stop valve requires continuous heat transfer from the heating medium with smallest temperature difference. This can be achieved only with infinite number of feed heating stages. But, in practice, it is not possible to have infinite heaters.

The number of feed water heaters used in the system is largely a problem of economic evaluation, in which the annual saving in fuel costs is balanced against annual fixed charges on heaters investment to determine how many heaters are required in a given installation for minimum generating costs. Five extraction points are generally used in practice. Some plants (using critical pressure) use as many as nine feed heaters.

4.1.8 Merits and Demerits of Gas Power Plant

Merits of gas power plant

- Natural gas is readily available.
- Setting-up cost can be reduced if the plant is installed near the fuel source.
- Less gas storage cost and space occupied is also less.
- Design and construction of gas power plant are simpler than those of a thermal.
- Power plant.
- Much smaller in size compared to the steam power plant of same capacity.
- Low operating cost.
- Less water consumption.
- Low maintenance cost.
- Can be started quickly and can be put on load in a very shorter time.
- No stand-by losses.
- Cheaper fuels like natural gas, kerosene and benzene are used.

Demerits of gas power plant

- Two-thirds of generated power is used for driving the compressor.
- Gas turbine has low thermal efficiency.
- Turbine blades have to be cooled by special cooling methods.
- Has starting problem.
- Efficient only in a combined cycle configuration.
- Temperature of combustion chamber is too high thereby resulting in a shorter life.

4.1.9 Solved Problems

Example 1: A simple ideal Brayton cycle with air as the working fluid has a pressure ratio of 11. The air enters the compressor at 300 K and the turbine at 1200 K. Accounting for the variation of the specific heats with temperature determine (*a*) the air temperature at the compressor and turbine exits, (*b*) the back work ratio, and (*c*) the thermal efficiency.

Assumptions: 1. Steady operating conditions exit. 2. The air-standard assumptions are applicable. 3. Kinetic and potential energy changes are negligible.

Analysis: (*a*) The air temperatures at the compressor and turbine exits are determined by applying the energy equation to the 4 processes involved in the Brayton cycle:

Process 1–2 (isentropic compression of an ideal gas):

Given: $T_1 = 300$ K

From steam table: $h_1 = 300.19$ kJ/kg

$Pr_1 = 1.386$ kN/m^2

$Pr_2 = (P_2/P_1)\ 8Pr_1 = (11)\ (1.386) = 15.284$

$\mathbf{T_2 = 579}$ **K** (at compressor exit)

$h_2 = 584.96$ kJ/kg

Process 3–4 (isentropic expansion of an ideal gas):

Given: $T_3 = 1200$ K

From steam table $h_3 = 1277.79$ kJ/kg

$Pr_3 = 238$ kN/m^2

$Pr_4 = (P_4/P_3)\ {*}Pr_3 = (1/11)\ (238) = 21.636$ kN/m^2

$$T_4 = \mathbf{648\ K} \text{ (at turbine exit)}$$
$$h_4 = 657.89 \text{ kJ/kg}$$

(*b*) To find the back work ratio, we need to find the work input to the compressor and the work output of the turbine:

$$\text{work input } w_{\text{comp, in}} = h_2 - h_1 = 584.96 - 300.19 = 284.77 \text{ kJ/kg}$$
$$\text{work output } w_{\text{turb, out}} = h_3 - h_4 = 1277.79 - 657.89 = 619.9 \text{ kJ/kg}$$

$$\text{Thus, Back work ratio (brw)} = \frac{w_{\text{comp, in}}}{w_{\text{turb, out}}} = 284.77/619.9 = 0.459$$

That is, **45.9 per cent** of the turbine work output is used just to drive the compressor.

(*c*) The thermal efficiency of the cycle is the ratio of the net power output to the total heat input:

$$\text{Heat input } q_{in} = h_3 - h_2 = 1277.79 - 584.96 = 692.83 \text{ kJ/kg}$$
$$\text{Net workdone } w_{net} = w_{out} - w_{in} = 619.9 - 284.77 = 335.13 \text{ kJ/kg}$$

$$\text{Thus, Thermal efficiency } \eta_{th} = \frac{W_{net}}{q_{in}} = 335.13/692.83 = \mathbf{0.484 \text{ or } 48.4\%}$$

Example 2: Consider an ideal gas-turbine cycle with two stages of compression and two stages of expansion. The pressure ratio across each stage of the compressor and turbine is 3. The air enters each stage of the compressor at 300 K and each stage of the turbine at 1200 K. Determine the back work ratio and the thermal efficiency of the cycle, assuming (*a*) no regenerator is used and (*b*) a regenerator with 75 per cent effectiveness is used. Use constant specific heats at room temperature.

Note: In the previous example the pressure ratio is 11 versus 3 in this example. Gas-turbine plants without intercooling, reheating, and regeneration operate more efficiently at higher pressure ratios. Gas-turbine plants incorporating intercooling, reheating, and regeneration operate more efficiently at lower pressure ratios. Had the plant in the previous example operated at a pressure ratio of 3 as in the current example, its back work ratio would have been 33.5% and the thermal efficiency would have been 25.4%.

Assumptions: 1. Steady operating conditions exits. 2. The air-standard assumptions are applicable. 3. Kinetic and potential energy changes are negligible.

Analysis: For two-stage compression and expansion, the work input is minimized and the work output is maximized when both stages of the compressor and the turbine have the same pressure ratio. Thus,

$$P_2/P_1 = P_4/P_3 = \sqrt{3} = 1.73 \text{ and } P_6/P_7 = P_8/P_9 = \sqrt{3} = 1.73$$

Air enters each stage of the compressor at the same temperature, and each stage has the same adiabatic efficiency (100 per cent in this case). Therefore, the temperature (and enthalpy) of the air at the exit of each compression stage will be the same. A similar argument can be given for the turbine. Thus,

At inlets: $T_1 = T_3, h_1 = h_3$ and $T_6 = T_8, h_6 = h_8$

At exits: $T_2 = T_4, h_2 = h_4$ and $T_7 = T_9, h_7 = h_9$

Under these conditions, the work input to each stage of the compressor will be the same, and so will the work output from each stage of the turbine.

(*a*) In the absence of any regeneration, the back work ratio and the thermal efficiency are determined as follows:

Given: $T_1 = 300$ K

From Steam Table: $h_1 = 300.19$ kJ/kg

$$Pr_1 = 1.386 \text{ kN/m}^2$$

$$Pr_2 = (P_2/P_1) * Pr_1 = (\sqrt{3})(1.386) = 2.401$$

$$T_2 = 351 \text{ K kN/m}^2$$

$$H_2 = 351.39 \text{ kJ/kg}$$

Given: $T_6 = 1200$ K

From Steam Table: $h_6 = 1277.79$ kJ/kg

$$Pr_6 = 238 \text{ kN/m}^2$$

$$Pr_7 = (P_7/P_6) * Pr_6 = (1/\sqrt{3})(238) = 137.4$$

$$T_7 = 1048 \text{ K}$$

$$H_7 = 1100.75 \text{ kJ/kg}$$

Then $w_{\text{comp, in}} = 2*(w_{\text{comp, in}}, 1) = 2*(h_2 - h_1) = 2*(351.39 - 300.19) = 102.4$ kJ/kg

$$w_{\text{turb, out}} = 2*(w_{\text{turb, out}}, 1) = 2*(h_6 - h_7) = 2*(1277.79 - 1100.75) = 354.08 \text{ kJ/kg}$$

$$w_{\text{net}} = w_{\text{turb, out}} - w_{\text{comp, in}} = 354.08 - 102.4 = 251.68 \text{ kJ/kg}$$

$$q_{\text{in}} = q_{\text{primary}} + q_{\text{reheat}} = (h_6 - h_2) + (h_6 - h_7)$$

$$= (1277.79 - 351.39) + (1277.79 - 1100.75) = 1103.44 \text{ kJ/kg}$$

Thus, Work ratio $= \dfrac{w_{\text{comp, in}}}{w_{\text{turb,out}}} = 102.4/354.08 = \mathbf{0.289 \text{ or } 28.9\%}$

and Thermal efficiency $\eta_{\text{th}} = \dfrac{w_{\text{net, in}}}{q_{\text{in}}} = 251.68/1103.44 = \mathbf{0.228 \text{ or } 22.8\%}$

A comparison with the previous example from single stage compression and expansion reveals that multistage compression with intercooling and multistage expansion with reheating significantly improves the back work ratio, but also significantly hurts the thermal efficiency. Therefore, intercooling and reheating are not recommended in gas-turbine power plants unless they are accompanied by regeneration.

(*b*) The addition of a regenerator does not affect the compressor work and the turbine work. Therefore, the network output and the back work ratio of an ideal gas-turbine cycle will be identical whether there is a regenerator or not. A regenerator, however, reduces the hot exhaust gases leaving the turbine. In an ideal regenerator, the compressed air is heated to the turbine exit temperature T_9 before it enters the combustion chamber. Thus, under standard air assumptions, $h_5 = h_7 = h_9$. Here, the regenerator is of 75 per cent effective so Pr_7 is reduced by 25%. Thus, the heat input and the thermal efficiency in this case are:

$$Pr_7 = (P_7/P_6)*Pr_6*.75 = (1/\sqrt{3})(238)(.75) = 103.057$$

From Steam Table: $T_5 = 975$ K

$$H_5 = 1017.32 \text{ kJ/kg}$$

Heat input $q_{\text{in}} = q_{\text{primary}} + q_{\text{reheat}} = (h_6 - h_5) + (h_6 - h_7)$

$$= (1277.79 - 1017.32) + (1277.79 - 100.75) = 437.51 \text{ kJ/kg}$$

and Thermal efficiency $\eta_{\text{th}} = \dfrac{w_{\text{net}}}{q_{\text{in}}} = 251.68/437.51 = \mathbf{0.575 \text{ or } 57.5\%}$

That is, the thermal efficiency increases by approximately 10% as a result of regeneration compared the first example without intercooling, reheating, and regeneration. The power put into compression is reduced by 255.32 kJ/kg because of intercooling, reheating, and regeneration while the power output decreases by 83.45 kJ/kg because of the lower pressure ratio. If the gas flows through the cycle at 18.14 kg/s, the cycle uses 4632 kJ/s or kW less in compression and produces 1514 kW less power. Including intercooling, reheating, and regeneration is usually well worth the extra cost associated with the second stage. A power generation plant, in an ideal situation, is in production mode 24 hours a day for 365 days per year. This equates to 8,736 hours per year. Businesses pay an average cost of $0.04173/kWh. At this price, the power generation plant would realize $1,136,576 additional profits per year with the reduction of compression electricity required. Adding more stages (no matter how many) can increase the efficiency an additional 7.3 percentage at most and usually cannot be justified economically.

4.2 DIESEL ENGINE POWER PLANT

Introduction

Diesel power plants are divided into two main classes: stationary and mobile. Stationary diesel power plants use four-stroke diesel engines (less frequently, two-stroke diesel engines), with power ratings of 110, 220, 330, 440, and 735 kilowatts (kW). Stationary diesel power plants are classed as average in their power rating if the rating does not exceed 750 kW; large diesel power plants can have a power rating of 2,200 kW or more. The advantages of a diesel power plant are favorable economy of operation, stable operating characteristics, and an easy and quick start-up. The main disadvantage is the comparatively short interval between major overhauls. Diesel power plants are used mainly for servicing areas remote from transmission lines or areas where sources of water supply are limited and where the construction of a steam power plant or of a hydroelectric power plant is not feasible. Stationary diesels are usually equipped with synchronous generators.

The economic efficiency of a diesel power plant is improved considerably if the waste heat of the engine (55 to 60 per cent of total heat release in currently available engines) can be used for preheating of fuel and oil or for domestic heating within the power station building or adjacent premises. In diesel power plants with a high power rating (above 750 kW) the waste heat can be used in a heating system serving a whole block or a whole town area in proximity to the power station.

Automatic protection against exceeding maximum or minimum limits for the temperature of cooling water and oil, the oil pressure, and the rotational speed (rpm) is built into diesel power plants; protection is also provided in the event of a short circuit in the line. Three levels of automation for stationary diesels are used: automatic regulation of the rotational speed (rpm) and of the temperature of the cooling water and oil, along with automatic emergency signaling and protection in the event of a breakdown; automatic or remotely controlled start-up and shutdown of the diesel engines, an automatic check of conditions required for connecting load to the line, synchronization with other units and with the power system, and a load connection and load distribution with units operating in parallel; and automatic refilling of the feeder tanks for fuel, oil, and water and of the air feed vessels, an automatic (trickle) charging of start-up batteries and of batteries used in auxiliary operations, and automatic control of the auxiliary equipment.

Mobile diesel power plants are widely used in agriculture and forestry and by expeditions involved with geological exploration. In these applications, diesel power plants can be used as a source of electricity for energy or lighting networks; they can be used as the main, auxiliary, or standby power source. In transportation, diesel power plants are a basic power source (for instance, in diesel-electric locomotives and in diesel ships). In mobile diesel power plants, the high-speed diesels serve as prime movers. A mobile diesel power plant includes the diesel-electric unit itself, spare parts, instruments and accessories, a set of cables for making connections to the load, and fire-fighting equipment. Automated diesel power plants with a power rating up to 10 kW are often mounted on a single-axle truck trailer; power plants rated 20 kW or more are usually installed on two-axle, covered trailers. Such a mobile station comprises not only the diesel-electric unit but also the power distribution cabinet (or panel), a cabinet containing the automatic controls, the remote control console, heating and ventilation equipment, rectifiers, and the storage batteries that feed the automatic controls or automated systems.

The first mobile diesel power plants in the USSR were built in 1934 and were known as diesel trains. Such diesel trains have all the power plant equipment installed on platforms or in cars. The power ratings of diesel trains are 1, 2.5, 4.5, and 10 megawatts.

The electric part of the power plant of a diesel train consists of a synchronous generator delivering a voltage of 3 to 10 kilovolts, assembled or unitized compartments containing high-voltage leads (overhead leads or cables), distribution equipment for voltages of 230–380 volts (required for lighting and for auxiliary motors of the power plant), the storage battery, and operating power circuits and the battery charger.

4.3 FUEL SUPPLY SYSTEM IN DIESEL ENGINE

The layout of fuel supply system is diesel engine is shown in Figure 4.6.

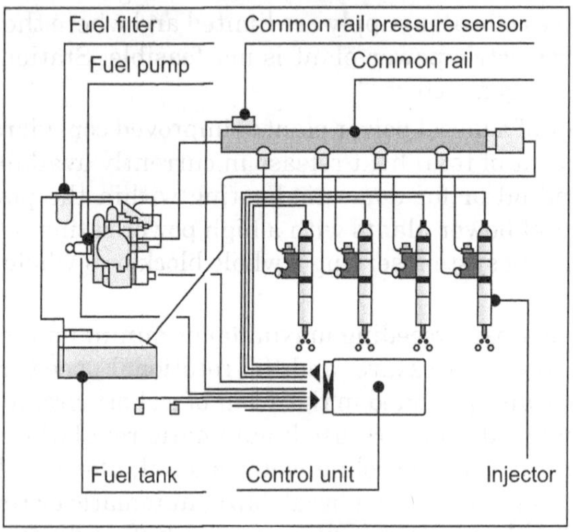

Fig. 4.6. Fuel Supply System

- It consists of a fuel tank, fuel feed pump, fuel filter, fuel injection pump and injector.
- Fuel is pumped to the fuel injection pump from fuel tank through fuel filter.
- From the fuel injection pump the fuel is delivered to the fuel nozzle or injectors.
- These injectors spray the fuel into the cylinder in the form of fire atomized spray.

Fuel Injection Pump

The fuel injection pump for diesel engine is shown in Figure 4.7.

- C.A.V. is the most commonly used fuel injection pump. The function of this pump is to supply the correct quantity of the fuel required for the working stroke at high pressure.

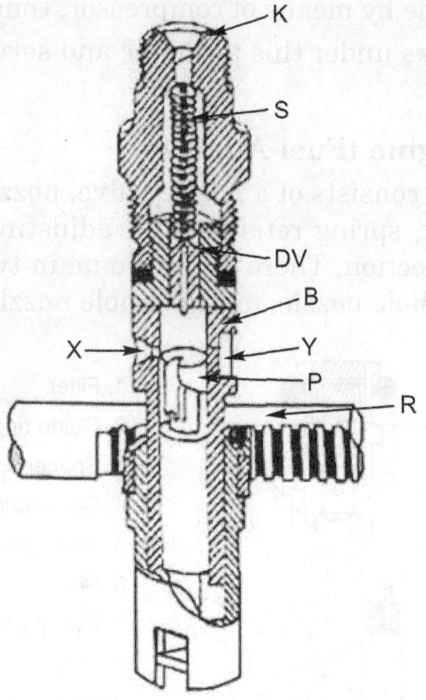

P = Plunger S = Spring
B = Barrel K = Passage
DV = Delivery valve R = Rack
X = Supply port Y = Spill port

Fig. 4.7. C.A.V Fuel Injection Pump

Description

- It consists of a plunger which reciprocates inside the barrel. There are so many plungers as the number of cylinders in the engine. The plunger has a vertical rectangular groove which extends from top to another helical groove cam drives the plunger. Fuel delivery valve is seated in its seat by the force of the spring.

Working Principle

- When the cam is not in action, the plunger is at the bottom of its stroke. Supply port and the spill port are uncovered. Fuel from the feed pump is forced into the barrel after filtration. When the cam pushes the plunger, it moves up and closes both the supply and spill ports. The fuel above the piston is compressed and high pressure is developed.
- Due to high pressure developed, the delivery valve is lifted off its seal and the fuel flows through the fuel passage to the atomizer.
- When the plunger moves further, helical groove connects the spill port to the top portion of the plunger. Consequently, there is a sudden pressure drop and the delivery valve is brought to its seat back under the action of the spring force.
- This cycle is repeated, rack and pinion arrangement which rotates the plunger inside the barrel controls the quantity of fuel.
- When the pressure falls, the nozzle valve is brought to its seat back under the spring force. If there is any leakage of fuel through the plunger of the nozzle valve, to a high temperature body some external work has to be done according to the second law of thermodynamics as shown.

- This external work is done by means of compressor, condenser etc.
- The machine, which works under this principle and serves the purpose of refrigeration, is called a refrigerator.

Fuel Injector of Diesel Engine (Fuel Atomizer)

The fuel injector Figure 4.8 consists of a nozzle valve, nozzle body, spring cap nut, spindle injector body, pressure spring, spring retainer nut, adjusting screw, lock nut for adjusting screw, fuel inlet, leak off connection. There are three main types of injection nozzles used in the atomizer. They are single hole nozzle, multiple hole nozzle and pintle nozzle.

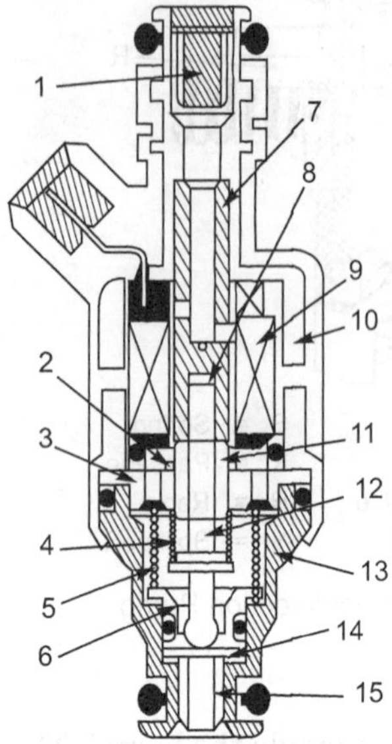

1. Filter
2. Guide ring
3. Spacer
4. Core spring
5. Seat spring
6. Seat
7. Pole piece
8. Stop
9. Solenoid coil
10. Solenoid body
11. Core ring
12. Core
13. Spray tip housing
14. Director or follower
15. Spray tip

Fig. 4.8. Fuel Injector

Working Principle: Fuel pump supplies the high pressure fuel into the fuel injector through the fuel passage. The nozzle valve is lifted off its seal against the spring force due to the fuel pressure. Then the fuel is injected into the combustion chamber in the form of fine spray. The fuel is atomised when the high pressure fuel is discharged through small sized holes of the nozzle atomiser.

4.4 TYPES OF DIESEL POWER PLANTS

Based on number of strokes the diesel engines can be classified as:

(*i*) Two stroke diesel engine (*ii*) Four stroke diesel engine

Based on orientation

(*i*) Horizontal diesel engine (*ii*) Vertical diesel engine

Depending on the number of cylinders

(*i*) Single cylinder (*ii*) Multi cylinder

and other type like naturally aspirated, supercharged, etc.

4.5 WORKING PRINCIPLE OF DIESEL POWER PLANT

The layout of Diesel power plant is shown in Figure 4.9. It consists of following main components.

1. **Diesel Engine:** It is the main component which develops the power to run the alternator connected to it to produce electrical power. In diesel engine, air admitted into the cylinder is compressed. At the end of the compression stroke, diesel is injected. It burns and the burning gases expand and do work on the piston. The reciprocating motion of the piston is converted into the rotary motion by crank.

   Since the crank shaft is directly coupled with the generator, the generator is directly run by the engine. The exhaust gases of the engine are then admitted to the atmosphere.

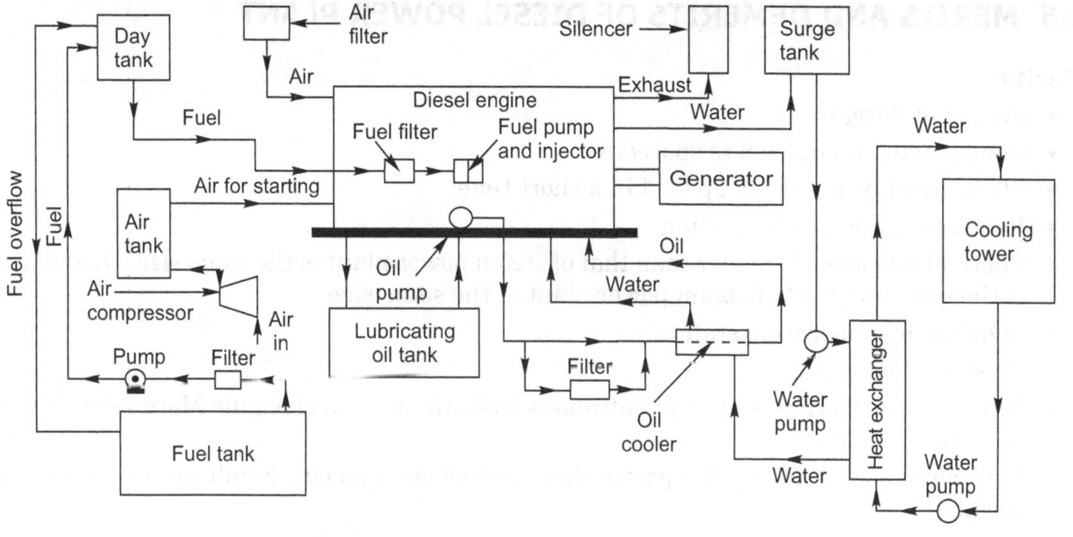

Fig. 4.9. Layout of Diesel Power Plant

2. **Air Intake System:** Air is drawn from the atmosphere during the suction stroke of the engine through an air filter which removes dust from the atmospheric air. A supercharger is used to increase the pressure and density of air before admission and this helps to increase the power output as more air-fuel mixture is burnt.
3. **Exhaust System:** Silencers are used to reduce the raise due to exhaust gases.
4. **Fuel System:** Fuel is stored in a large reservoir constructed below the ground floor. Pump draws diesel from storage tank and supplies it to the small tank through the filter. Day tank supplies the daily fuel need of the engine. The day tank supplies the fuel by gravity action, as it is placed high. Fuel injection pump injects the fuel into the engine after filtration. The fuel pump also regulates the fuel according to the load on the engine.
5. **Cooling System:** It is classified into air cooled and water cooled systems. The water cooling system consists of water circulating pumps, cooling towers and water filtration plants. To keep the temperature of the engine at reasonable level, water is circulated inside the engine in water jackets around the cylinder, piston, combustion chamber, etc. Hot water leaving the jacket is sent to the heat exchanger where the hot water is cooled by the raw water. This heated raw water is then cooled in the cooling tower and filtered in the filtration tank and recirculated again.
6. **Lubrication System:** It includes lubricating oil tank, oil pump and cooler. Lubrication is used to reduce the friction and wear of the engine parts such as cylinder walls and

piston. Lubricating oil which gets heated due to the friction of moving parts is cooled by the cooling water which is common for the engine cooling, before recirculation.

7. **Engine Starting System:** There are three starting methods. They are by an auxiliary engine, by a electric motor and by compressor. It includes air compressor and starting air tank. The function of this system is to start the engine from cold by supplying the compressed air. Small diesel engines can be crank started but large diesel engines such as those used in power plants can be started only by compressed air.
8. **Governing System:** It is used to maintain the speed of the engine constant at all loads by adjusting the amount of fuel supplied to the engine.

4.6 MERITS AND DEMERITS OF DIESEL POWER PLANT

Merits:

- Simple in design
- Compact and occupies less space/area
- Starts quickly and picks up load in a short time
- Requires less water for cooling
- Thermal efficiency is higher than that of steam power plant of the same size. Overall cost is cheaper than that of steam power plant of the same size
- Requires less operating staff
- No stand-by losses
- It can deliver more power on a continuous basis than a petrol engine More reliable than other power plants
- Extremely useful during the power shortages or emergencies Small diesel engines are portable.

Demerits:

- High operating cost due to high diesel price
- Plant does not work efficiently under prolonged overload condition. Generates only small quantity of power
- Lubrication cost is high
- Maintenance charges are generally high
- Increased oil consumption
- It is noisy
- Pollutes the environment

4.7 APPLICATION OF DIESEL POWER PLANT

1. Used as peak load plants.
2. Suitable for mobile plants.
3. Used as standby units.
4. Used as emergency plant.
5. Used for small and nursery stations.
6. Used as starting stations.
7. Used in small capacity central station.

4.8 COMPARISON OF POWER PLANTS

(i) Comparison based on location selection, space required and setting-up cost

Power Plant	Location Selection	Space required	Setting-up cost
Steam/Thermal	Where there is ample supply of water and coal	Needs sufficient space for all equipment	Lower than hydro and nuclear
Hydro	Where large reservoirs can be formed by constructing dam	Needs very large space for constructing a dam	Very high because dam construction
Diesel	Can be located anywhere	Needs less space for all equipment	Lower compared to all other power plants
Nuclear	Where there is enough supply of water, but must be far away from populated areas	Needs minimum space for all equipment	Highest due to complex nuclear reactors

(ii) Comparison based on running cost, efficiency and stand-by losses

Power Plant	Running cost	Efficiency	Stand-by losses
Steam/Thermal	Higher than hydro and nuclear	About 25%	Stand-by losses are heavy due to boiler remaining in operation even after the turbine is switched off
Hydro	Practically nil because no fuel is needed	About 85%	This plant has no stand-by losses
Diesel	Highest due to diesel prices	About 35%	This plant has less stand-by losses
Nuclear	Low	About 55%	This plant has less stand-by losses

(iii) Comparison based on cost of fuel, clean emissions and limitation of source

Power Plant	Cost of fuel	Clean emissions	Limitation of source
Steam/Thermal	Maximum due to constant need of coal and transportation	Highest polluting emissions	Most limited fuel reserves
Hydro	No fuel cost	No emission	Depends on water source
Diesel	Higher than nuclear power plant	Cleaner emissions compared to steam and nuclear power plants	Limited fuel reserves
Nuclear	Minimum duc to small quantities of fuel required	Cleaner emissions compared to steam power plant	Sufficient fuel reserves

(*iv*) Comparison based on transmission and distribution cost, maintenance cost and starting time required

Power Plant	Transmission and distribution cost	Maintenance cost	Starting time required
Steam/Thermal	Low, since plant is closer to load centre	High due to skilled operating staff required	Longest starting time
Hydro	High, since plant is far away from load centre	Low	Shortest starting time
Diesel	Low, since plant is at load centre	Low	Short starting time
Nuclear	High, since plant is far away from load centre	Highest due to expert operating staff required	Long starting time

REVIEW QUESTIONS

Two Marks Questions

1. What is meant by gas turbine power plant?
2. What are the major components of gas turbine power plant?
3. What are the types of gas turbine power plant?
4. What are the working fluids in gas turbine?
5. What are the major advantages of gas turbine?
6. How the gas turbine blades are cooled?
7. How thermal efficiency of gas turbine plant can be increased?
8. How gas turbines are classified?
9. Sketch the schematic arrangement of open cycle gas turbine plant and name the components.
10. Mention the various process of the Brayton cycle.
11. Draw the *p-V* and *T-S* diagram of Brayton cycle.
12. Differentiate gas and steam turbine.
13. Differentiate Gas turbine and IC Engine.
14. Define the following:
    (*i*) Intercooling (*ii*) Preheating
15. List the merits and demerits of gas turbine power plant.
16. What is the principle of operation of simple jet propulsion system?
17. Why is the maximum cycle temperature of gas turbine plant much lower than that of diesel power plant?
18. What is meant by diesel power plant?
19. What are all the modifications carried out in a brayton cycle? Why?
20. Sketch the limited pressure cycle on *P-v* and *T-S* diagram and name various processes.
21. What is the use of regenerator?
22. What are the effects of introducing regeneration in the basic gas turbine cycle?
23. When will the intercooler is provided between the two compressors?
24. When reheater is employed in gas turbine cycle?
25. What are the effects of reheat cycle?
26. What is fuel injection system in diesel engine?

27. How the fuel injection systems are classified?
28. What are all the performance parameters in IC Engine?
29. What are the methods of lubrication system in IC engine?
30. What are the different types of governing system in IC engine?
31. What are the methods of starting system in large and medium size engines?
32. What is meant by reheating combustion chamber?
33. Give the properties of fuel for gas turbine?
34. List out the inherent advantages of the combined power cycles.
35. How diesel power plants are classified?
36. What are the major components of diesel power plant?
37. What are the methods of cooling system?
38. What are the functions of lubricating system?
39. What are the starting methods in diesel engine?
40. What are the merits and demerits of diesel power plant?
41. What are the applications of diesel power plant?

Sixteen Marks Questions

1. Explain the working principle of closed cycle gas turbine power plant with neat diagram.
2. Explain the working principle of open cycle gas turbine power plant with neat diagram.
3. Explain the working principle of gas turbine with intercooling.
4. Describe with suitable diagram:
   (*i*) Open cycle gas turbine (*ii*) Closed cycle gas turbine
5. Explain the effect of intercooling reheating and regeneration with *P-V* and *T-S* diagram.
6. Explain the merits and demerits of gas power plant.
7. Explain the fuel injection system of diesel engine with a neat sketch.
8. Explain the working principle of fuel injection pump with a neat sketch.
9. Discuss in detail about the lubrication system and cooling system of diesel engines.
10. Explain the working principle of diesel power plant with neat diagram.
11. Explain the merits and demerits of diesel power plant.
12. Compare the power plants with:
    (*i*) Location, required and setting up cost.
    (*ii*) Running cost, efficacy and stand by loss.
    (*iii*) Cost of fuel, clean emission and limitation of sources.
    (*iv*) Transmission and distribution cost, maintenance cost and starting time required.

5 Non-Conventional Power Generating Systems

5.1 SOLAR POWER PLANT

In the solar power plant, solar energy is used to generate electricity. Sunrays are focused using concave reflectors on to copper tubes filled with water and painted black outside. The water in the tubes then boils and become steam. This steam is used to drive steam turbine, which in turn causes the generator to work for producing electricity.

Many power plants today use fossil fuels as a heat source to boil water. The steam from the boiling water rotates a large turbine, which activates a generator that produces electricity. However, a new generation of power plants, with concentrating solar power systems, uses the sun as a heat source. There are three main types of concentrating solar power systems: parabolic-trough, dish/engine, and power tower.

Parabolic-trough systems concentrate the sun's energy through long rectangular, curved (U-shaped) mirrors. The mirrors are tilted toward the sun, focusing sunlight on a pipe that runs down the center of the trough. This heats the oil flowing through the pipe. The hot oil then is used to boil water in a conventional steam generator to produce electricity.

A dish/engine system uses a mirrored dish (similar to a very large satellite dish). The dishshaped surface collects and concentrates the sun's heat onto a receiver, which absorbs the heat and transfers it to fluid within the engine. The heat causes the fluid to expand against a piston or turbine to produce mechanical power. The mechanical power is then used to run a generator or alternator to produce electricity.

A power tower system uses a large field of mirrors to concentrate sunlight onto the top of a tower, where a receiver sits. This heats molten salt flowing through the receiver. Then, the salt's heat is used to generate electricity through a conventional steam generator. Molten salt retains heat efficiently, so it can be stored for days before being converted into electricity. That means electricity can be produced on cloudy days or even several hours after sunset.

Concentration Ratio: It is defined as the ratio between the aperture area and the receiver/absorber area of the collector.

'Solar Power Tower' Power Plant the first is the 'Solar Power Tower' design which uses Thousands of sun-tracking reflectors or heliostats to direct and concentrate solar radiation onto a boiler located atop a tower. The temperature in the boiler rises to 500–7000°C and the steam raised can be used to drive a turbine, which in turn drives an electricity producing turbine. There are also called central Receiver Solar Power Plants.

It can be divided into solar plant and conventional steam power plant. The flow diagram is given in Figure 5.1.

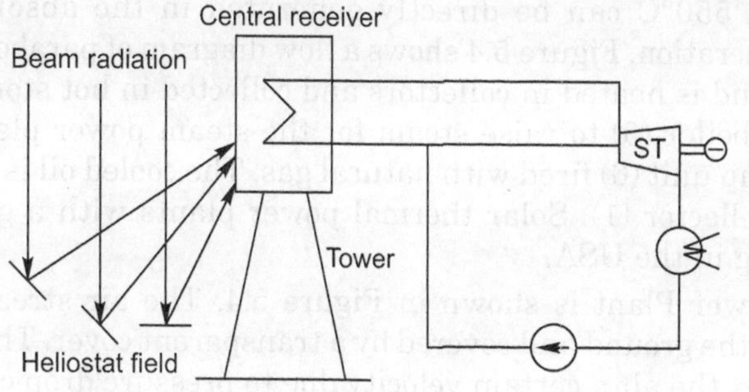

Fig. 5.1. Central Receiver Solar Power Plant

A heliostat field consists of a large number of flat mirrors of 25 to 150 m^2 area which reflects the beam radiations onto a central receiver mounted on a tower. Each mirror is tracked on two axis. The absorber surface temperature may be 400 to 1000°C. The concentration ratio (total mirror area divided by receiver area) may be 1500. Steam, air or liquid metal may be used as working fluid. Steam is raised for the conventional steam power plant.

'Distributed (Parabolic) Collector System' Power Plant. The second type is the distributed collector system. It is also called solar farm power plant as a number of solar modules consisting of parabolic trough solar collectors are interconnected. This system uses a series of specially designed 'Trough' collectors which have an absorber tube running along their length. Large arrays of these collectors are coupled to provide high temperature water for driving a steam turbine. Such power stations can produce many megawatts (MW) of electricity, but are confined to areas where there is sample solar insulation.

Every module consists of a collector as shown in Figures 5.2 and 5.3. It is rotated about one axis by a sun tracking mechanism. Thermo-oil is mostly used as heating fluid as it has very high boiling Beam point. Water/steam working fluid can also be used. The tubes have evacuated

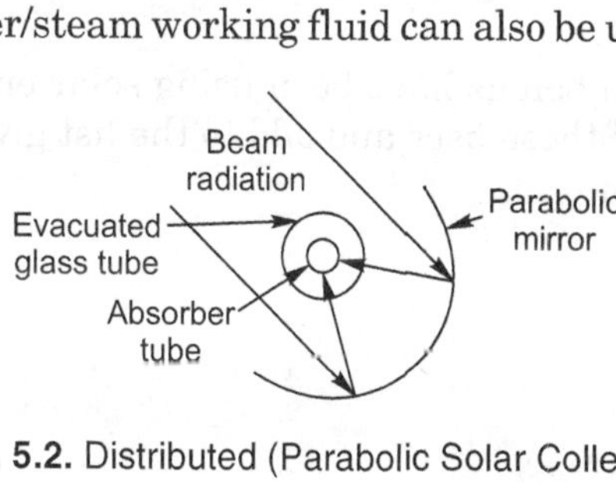

Fig. 5.2. Distributed (Parabolic Solar Collector)

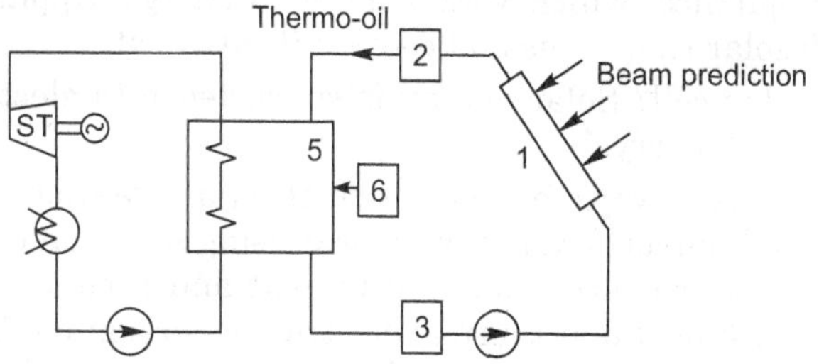

Fig. 5.3. Distributed (Parabolic) Trough Solar Power Plant

glass enclosure to reduce the losses. The concentration ratio is between 40 and 100. The maximum oil temperature is limited to 400°C as oil degrades above this temperature.

Alternately steam at 550°C can be directly generated in the absorber tube. These are commercially under operation. Figure 5.4 shows a flow diagram of parabolic trough solar power plant. The working fluid is heated in collectors and collected in hot storage tank (2). The hot thermo-oil is used in boiler (5) to raise steam for the steam power plant. The boiler also is provided with a back-up unit (6) fired with natural gas. The cooled oil is stored in tank (3) and pumped (4) back to collector (1). Solar thermal power plants with a generating capacity of 80 MW are functioning in the USA.

Solar Chimney Power Plant is shown in Figure 5.4. The air stream is heated by solar radiation absorbed by the ground and covered by a transparent cover. The hot air flow through or chimney which gives the air a certain velocity due to pressure drop caused by the chimney effect. The hot air flows through an air turbine to generate power.

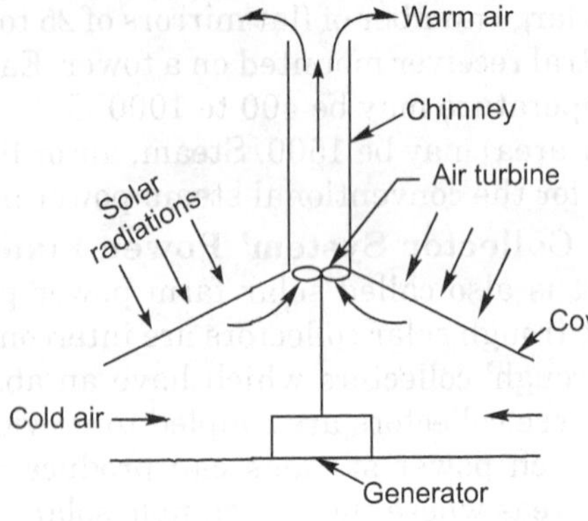

Fig. 5.4. Chimney Solar Power Plant

Solar Energy Storage

It is well known that human beings have been using solar energy for different uses, from ancient days. Find examples of these uses and add to the list given below.

1. To get salt from sea water
2. To dry wet clothes
3. To dry firewood
4. To dry cereals
5. To dry fish
6. To dry leather

We now use several appliances which work using solar energy. Appliances like solar cooker and solar heater absorb solar radiations and convert it into heat.

Then what about a solar cell? Solar energy is converted into electrical energy and it is directly used or stored in a battery.

There are eight possible pathways for conversion of solar radiation to useful energy. Solar thermal conversion method converts radiation to heat using solar flat collectors. Solar thermo chemical conversion method converts radiation to heat and produce steam then to kinetic energy using a pump or turbine. Solar thermal electric conversion method converts radiation to steam and to kinetic and electrical energy through a turbine and generator to electrical energy. The above route through a further electrolysis process gives chemical energy (H_2 fuel). A high temperature catalytic conversion process produces chemical energy (H_2 fuel) directly. Photovoltaic conversion of solar radiation gives direct electrical energy. Photosynthesis process produces chemical energy directly from radiation. Chemical energy (H_2 fuel) is directly produced from solar radiation using the electricity produced by the photovoltaic method. A few of these methods are dealt in detail further.

Commercial and industrial buildings may use the same solar technologies photovoltaic, passive heating, day lighting, and water heating that are used for residential buildings. These non-residential buildings can also use solar energy technologies that would be impractical for a home. These technologies include ventilation air preheating, solar process heating and solar cooling.

Many large buildings need ventilated air to maintain indoor air quality. In cold climates, heating this air can use large amounts of energy. A solar ventilation system can preheat the air, saving both energy and money. This type of system typically uses a transpired collector, which consists of a thin, black metal panel mounted on a south-facing wall to absorb the sun's heat. Air passes through the many small holes in the panel. A space behind the perforated wall allows the air streams from the holes to mix together. The heated air is then sucked out from the top of the space into the ventilation system.

Solar process heating systems are designed to provide large quantities of hot water or space heating for non-residential buildings. A typical system includes solar collectors that work along with a pump, a heat exchanger, and/or one or more large storage tanks. The two main types of solar collectors used an evacuated tube collector and a parabolic trough collector can operate at high temperatures with high efficiency. An evacuated-tube collector is a shallow box full of many glass, double-walled tubes and reflectors to heat the fluid inside the tubes. A vacuum between the two walls insulates the inner tube, holding in the heat. Parabolic troughs are long, rectangular, curved (U-shaped) mirrors tilted to focus sunlight on a tube, which runs down the center of the trough. This heats the fluid within the tube.

The heat from a solar collector can also be used to cool a building. It may seem impossible to use heat to cool a building, but it makes more sense if you just think of the solar heat as an energy source. Your familiar home air conditioner uses an energy source, electricity, to create cool air. Solar absorption coolers use a similar approach, combined with some very complex chemistry tricks, to create cool air from solar energy. Solar energy can also be used with evaporative coolers (also called "swamp coolers") to extend their usefulness to more humid climates, using another chemistry trick called desiccant cooling.

Space Heating

In colder areas of the world (including high altitude areas within the tropics) space heating is often required during the winter months. Vast quantities of energy can be used to achieve this. If buildings are carefully designed to take full advantage of the solar insolation which they receive then much of the heating requirement can be met by solar gain alone. By incorporating certain simple design principles a new dwelling can be made to be fuel efficient and comfortable for habitation. The bulk of these technologies are architecture based and passive in nature. The use of building materials with a high thermal mass (which stores heat), good insulation and large glazed areas can increase a buildings capacity to capture and store heat from the sun. Many technologies exist to assist with diurnal heating needs but seasonal storage is more difficult and costly.

For passive solar design to be effective certain guidelines should be followed:

1. A building should have large areas of glazing facing the sun to maximize solar gain.
2. Features should be included to regulate heat intake to prevent the building from overheating.

3. A building should be of sufficient mass to allow heat storage for the required period.
4. Contain features which promote the even distribution of heat throughout the building.

One example of a simple passive space heating technology is the Trombe wall. A massive black painted wall has a double glazed skin to prevent captured heat from escaping. The wall is vented to allow the warm air to enter the room at high level and cool air to enter the cavity between the wall and the glazing. Heat stored during the wall during the day is radiated into the room during the night. This type of technology is useful in areas where the nights are cold but the days are warm and sunny.

Space Cooling

The majority of the world's developing countries, however, lies within the tropics and have little need of space heating. There is a demand, however, for space cooling. The majority of the world's warm-climate cultures have again developed traditional, simple, elegant techniques for cooling their dwellings, often using effects promoted by passive solar phenomenon. There are many methods for minimizing heat gain. These include sitting a building in shade or near water, using vegetation or landscaping to direct wind into the building, good town planning to optimize the prevailing wind and available shade. Buildings can be designed for a given climate domed roofs and thermally massive structures in hot arid climates, shuttered and shaded windows to prevent heat gain, open structure bamboo housing in warm, humid areas. In some countries dwellings are constructed underground and take advantage of the relatively low and stable temperature of the surrounding ground. There are as many options as there are people.

Recent Developments in Solar Power Plants

Solar Thermal Applications. The applications include water heating for domestic, commercial and industrial use, space heating and drying, solar distillation, solar cooling through absorption and adsorption cycles, solar water pumping and solar power generation.

Solar Photovoltaics. Photovoltaic (*PV*) or solar cells refers to the creation of voltage from light. A solar cell is a converter; it changes the light energy into electrical energy. A cell does not store any energy, so when the source of light (typically the sun) is removed, there is no electrical current from the cell. If electricity is needed in the night, a battery must be included in the circuit. There are many materials that can be used to make solar cells, but the most common is the element silicon. A typical solar cell is 3 to 6 inches in diameter and are now available in various shapes like circular, square, etc. The conversion processes occurs instantly whenever there is light falling on the surface of a cell. And the output of the cell is proportional to the input light.

5.2 WIND POWER GENERATION

You usually stand in an open space to enjoy the wind. You know how wind originates. Moving air is wind. Since the wind has velocity it has kinetic energy. This is the energy of the wind. We shall see how the kinetic energy of the wind can be used to produce electricity. For that, we can use windmills. Windmills are devices, which work on wind. How the kinetic energy of the wind is made use of in windmills shall be looked into. We shall examine the working of a windmill. The important part of a windmill Figure 5.5 (*a*) is a structure with large leaves, fixed at the top of a high tower. Figure 5.5 (*b*) illustrates the major component of wind energy system.

What will happen when wind blows on these leaves? You may have seen paper fans available at festival places, rotating when the wind blows. In a similar manner the speed of leaves changes with the speed of the wind. What happens if the rotation of the windmill is given to the rotor of a generator? Rotor also rotates. Then electricity is obtained from the generator. What happens if the windmill is connected to a water pump? As the leaves of the windmill rotate pump works pumping out water. Figure 5.5 (*c*) shows the layout of single wind energy system.

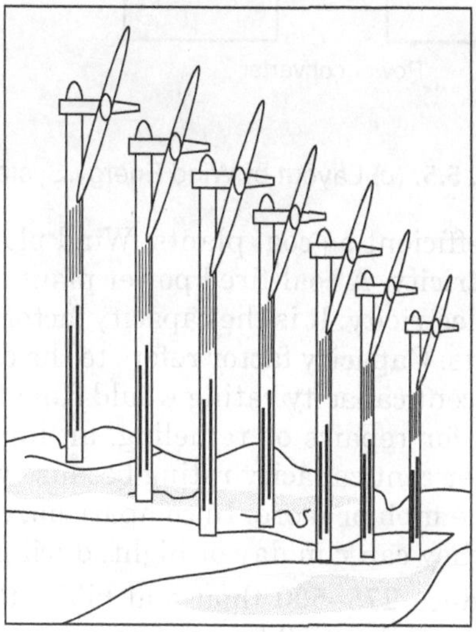

Fig. 5.5. (*a*) Windmill Farm

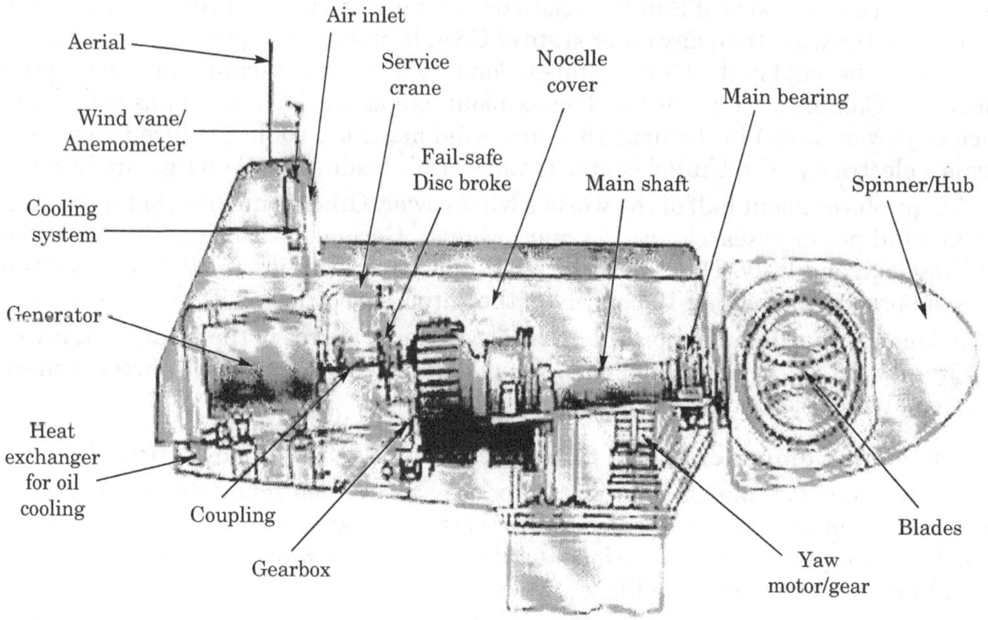

Fig. 5.5. (*b*) Major components of the Wind Energy System

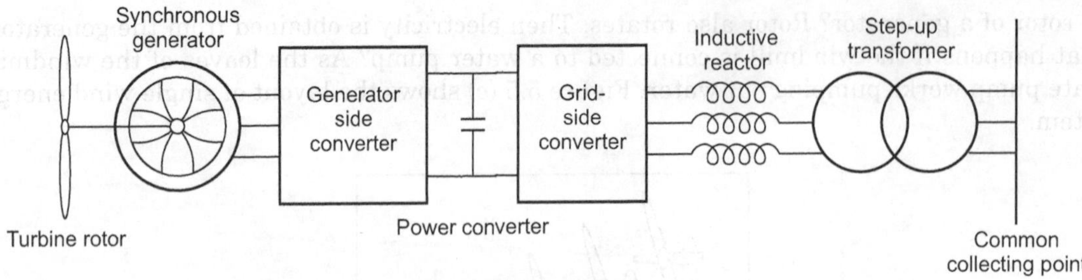

Fig. 5.5. (*c*) Layout of Wind Energy System

Wind machines are just as efficient as coal plants. Wind plants convert 30 per cent of the wind's kinetic energy into electricity. A coal-fired power plant converts about 30–35 per cent of the heat energy in coal into electricity. It is the capacity factor of wind plants that puts them a step behind other power plants. Capacity factor refers to the capability of a plant to produce energy. A plant with a 100 per cent capacity rating would run all day, every day at full power. There would be no down time for repairs or refueling, an impossible dream for any plant. Wind plants have about a 25 per cent capacity rating because wind machines only run when the wind is blowing around nine mph or more. In comparison, coal plants typically have a 75 per cent capacity rating since they can run day or night, during any season of the year.

One wind machine can produce 275–500 thousand kilowatt-hours (kWh) of electricity a year. That is enough electricity for about 50 homes per year. In this country, wind machines produce about three billion kWh of energy a year. Wind energy provides 0.12% of the nation's electricity, a very small amount. Still, that is enough electricity more than 300,000 households, as many as in a city the size of San Francisco or Washington, D.C. California produces more electricity from the wind than any other state of USA. It produces 98 per cent of the electricity generated from the wind in the United States. Some 16,000 wind machines produce more than one per cent of California's electricity. (This is about half as much electricity as is produced by one nuclear power plant.) In the next 15 years, wind machines could produce five per cent of California's electricity. The United States is the world's leading wind energy producer.

The U.S. produces about half of the world's wind power. Other countries that have invested heavily in wind power research are Denmark, Japan, Germany, Sweden, The Netherlands, United Kingdom, and Italy. The American Wind Energy Association (AWEA) estimates wind energy could produce more than 10 per cent of the nation's electricity within the next 30 years.

So, wind energy may be an important alternative energy source in the future, but it will not be the sole answer to our energy problems. We will still need other energy sources to meet our growing demand for electricity.

Energy-efficient Components and Subsystems for High System Reliability

Power semiconductors play a key role to produce energy from renewable sources. In wind turbines, power semiconductors are used to convert power and to couple the generator with the grid. They are also built into various auxiliary drives such as yaw drives, pitch drives, pumps and into protection circuits like crowbars.

Wind power converters control a number of vital functions and applications and therefore require power semiconductors of the highest quality standards. This applies in particular to offshore wind converters which operate in exceptionally harsh environments exposed to salt, humidity etc. Rapid growth is projected for the offshore segment.

Wind energy turbines must also be designed to deliver maximum levels of availability in order to contribute to grid stability. This applies not only to the converter, but also to the various auxiliary drives mounted in different positions. Grid stability therefore depends on power semiconductor assemblies offering dynamic capabilities, outstanding functionality and superior reliability.

Fixed-speed Generators

Fixed-speed designs are suited to generators up to 400 kW. New designs are usually based on semiconductor solutions to fulfill grid code requirements. Nevertheless, infineon delivers the full range of grid coupling components, with thyristors and other bipolar semiconductors available as modules, discs and stacks.

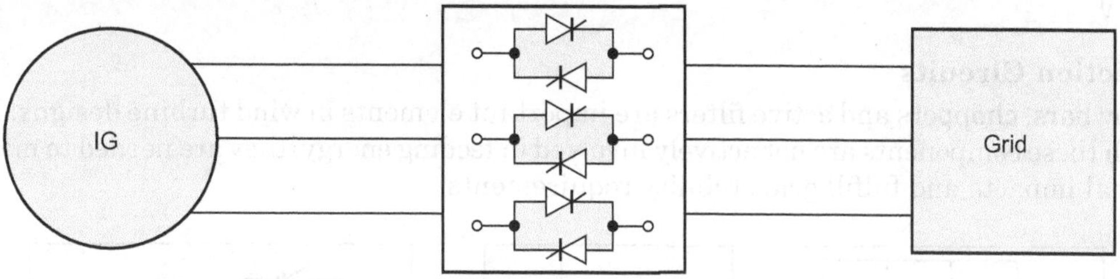

Doubly Fed Induction Generator (DFIG)

This speed-adjustable design is typically deployed in the power range between 400 kW and 2.500 kW. To control the full load, only up to one-third of the energy needs to be converted over power semiconductors in both directions. Infineon's IGBTs support optimum performance even at the limits of the operating range. High quality design, low voltage ride through (LVRT) capability and very low output frequencies meet high availability needs.

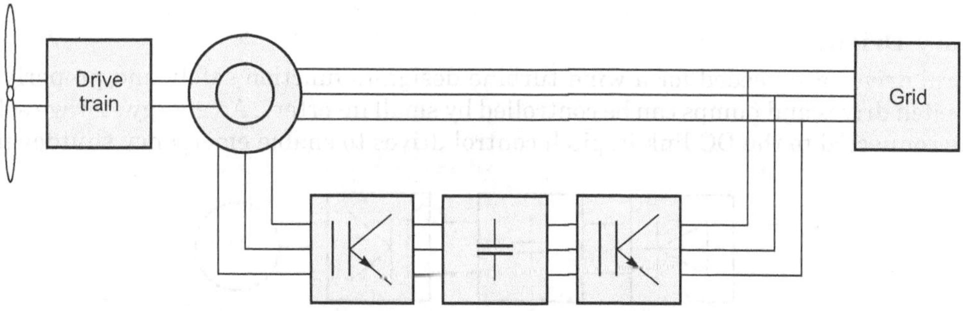

Permanent Magnet (PM) Generators

Full converters for low-, medium- and high-speed generators provide maximum flexibility to meet LVRT and other grid stability requirements. High efficiency is mandatory and an active front-end inverter adapts variable power and frequency. This design reduces inductive component effort. Infineon's IGBTs enable modular and scalable system designs.

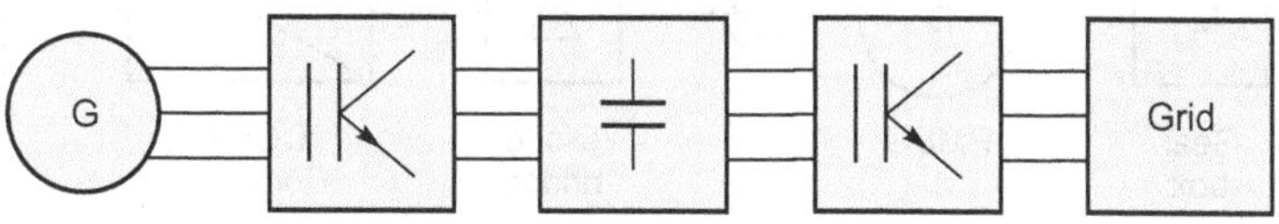

Electrical Excited (EE) Generators

An EE generator feeds the inverter via a bipolar rectifier, with the generator voltage controlled by excitation. The grid inverter is controlled by an IGBT as the effect of excitation. This design is an established solution already successfully deployed in modern systems.

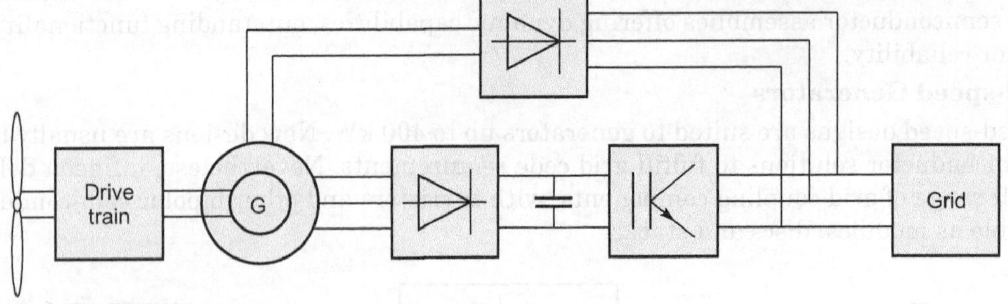

Protection Circuits

Crowbars, choppers and active filters are important elements in wind turbine designs. Even though these components are not actively involved in feeding energy, they are needed to manage external impacts and fulfill grid stability requirements.

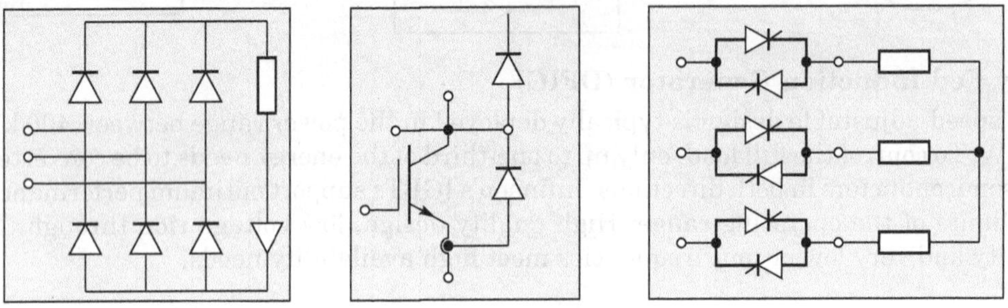

Auxiliary Drives

Several drives are needed for a wind turbine design to function safely and properly. Yaw drives, pitch drives and pumps can be controlled by small inverters. An energy storage solution must be connected to the DC link in pitch control drives to enable emergency shutdown.

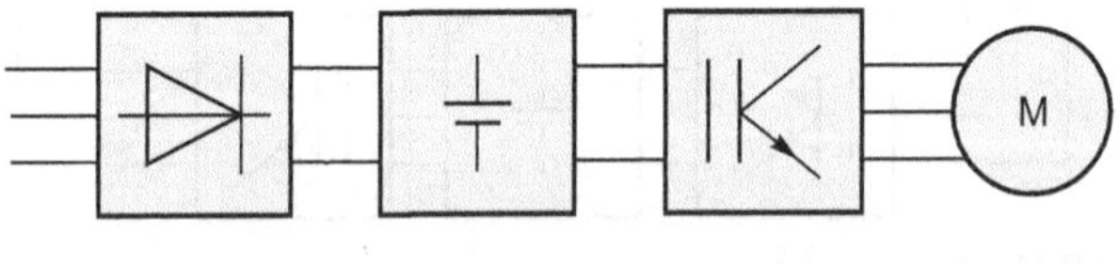

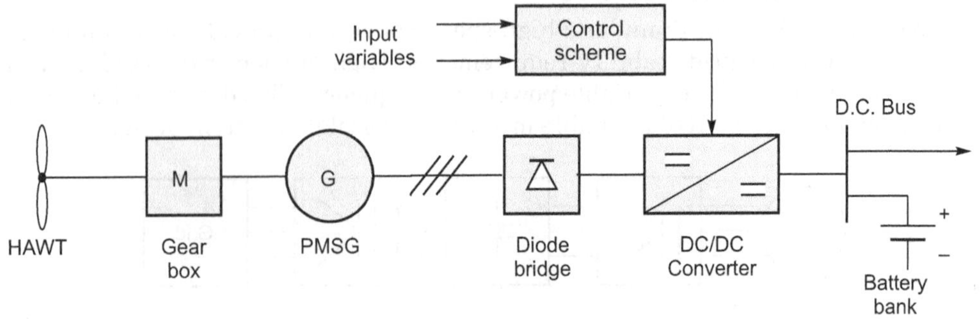

5.3 TIDAL POWER GENERATION

The generation of electricity using tidal power is basically the transformation of tidal power found in tidal motion of water in seas and oceans into electrical energy. This is done using a very basic idea involving the use of a barrage or small dam built at the entrance of a bay where tides are known to reach very high levels of variation.

This barrage energy will be converted into rotational kinetic energy that will rotate turbines giving electrical energy. Figure 5.6 shows the process in very simple terms.

- Tidal stream systems make use of the kinetic energy from the moving water currents to power turbines, in a similar way to wind mills use moving air. This method is gaining in popularity because of the lower cost and lower ecological impact.
- Barrages make use of the potential energy from the difference in height (or head) between high and low tided. Barrages suffer from the problems of very high civil infrastructure costs, few viable sites globally and environmental issues.

Tidal energy can be created in there different ways:

1. Tidal Barrages
2. Tidal Fences
3. Tidal Turbine

Tidal Barrages

A tidal barrage is basically a huge dam that is built across a river or estuary. Tunnels are built throughout the dam, and when the tide rises and falls it turns a turbine that generates power.

Tidal Fences

Tidal fences are giant turnstiles that have vertical axis turbines mounted on a fence. The water passes through the fence and powers the turbines.

Tidal Turbine

Tidal turbines (Figure 5.6) are the latest water power technology used in tidal areas. They look like wind turbines that are lined together underwater. They are used to capture strong tidal flows, so they to be built extra strong.

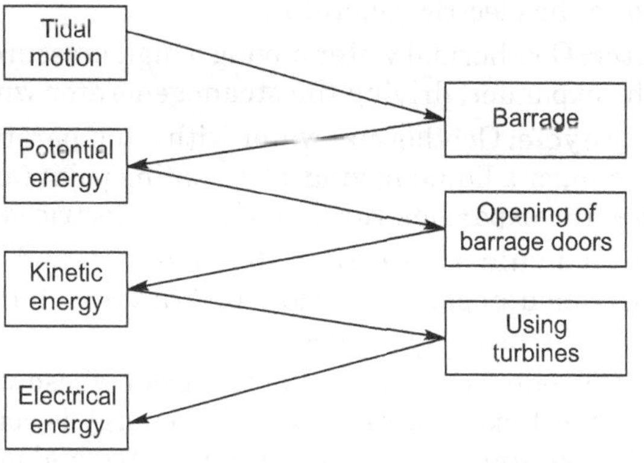

Fig. 5.6. Transformation of Tidal Energy to Electric Energy

Tidal barrages, undersea tidal turbines – like wind turbines but driven by the sea and a variety of machines harnessing undersea currents are under development. Unlike wind and waves tidal currents are entirely predictable.

5.4 GEOTHERMAL POWER PLANT

Geothermal comes from the Greek words geo, meaning earth, and thermo, meaning heat. Geothermal energy is the residual thermal energy in the Earth left over from the planet's origins, usually occurring in deep reservoirs of hot fluids within the Earth's crust.

It is the challenge of geothermal scientists and engineers to locate these reservoirs of hot fluids, design means to bring the fluids to the surface, process them in a suitable power plant to generate electricity and then dispose of the spent fluids in an acceptable manner. Three different types of power plants—dry steam, flash, and binary—are used, to generate electricity from geothermal energy, depending on temperature, depth, and quality of the water and steam in the area.

In all cases the condensed steam and remaining geothermal fluid is injected back into the ground to pick up more heat. In some locations, the natural supply of water producing steam from the hot underground magma deposits has been exhausted and processed waste water is injected to replenish the supply.

Geothermal Technologies

Generation of electricity from geothermal energy typically involves drilling wells into naturally-heated subterranean aquifers and then expanding a hot fluid—either the fluid produced directly from those wells or a secondary fluid heated by the directly-produced fluid—through a turbo expander that drives an electric generator.

While commercial geothermal power projects are now successfully operating in five states, further development of the nation's geothermal resources would benefit from improved technology for drilling wells into geothermal aquifers, producing fluids from them, and generating electricity from those fluids.

Working Principle

Geothermal power plants (GPPs) (Fig. 5.7) use the heat of the geothermal water or geothermal steam to generate electricity. Based on the state and the temperature of the geothermal water used, there are a number of types of geothermal power plants. The basic types are as follows:

- **GPPs with overheated steam:** The steam coming out of the well following the water separation drives the steam turbine with the generator or the steam is fed to the steam generator (heat exchanger), where the surface water produced steam drives the steam generator connected to the electric generator.
- **GPPs with hot water:** Geothermal water having a high pressure and temperature turns into wet steam in the expander, driving the steam generator with the generator.
- **GPPs with a binary cycle:** Geothermal water with a temperature of above some 130°C heats in the heat exchanger a liquid having a low boiling point (ammonia, isobutane), the steam of which drives the expansion turbine with the electric generator.
- **New sources have as a rule a pair of wells:** Producing and re-injection well. Cooled waters along with detrimental gases and salts return through the re-injection well back into the ground for environmental reasons.

Origin of Geothermal (GT) energy is hot earth core, from whose warmth escapes through rifts in rock. Concerning almost exhaustless stocks, GT energy is considered to be revivable power source. According to potential of supply exist: reinserted and no reinserted GT sources. Reinserted sources claim only cutting bore and cooled water is drained to watercourses. No reinserted sources include additional re-injection bore, which draw cooled water back to underground reservoir to complement water supply.

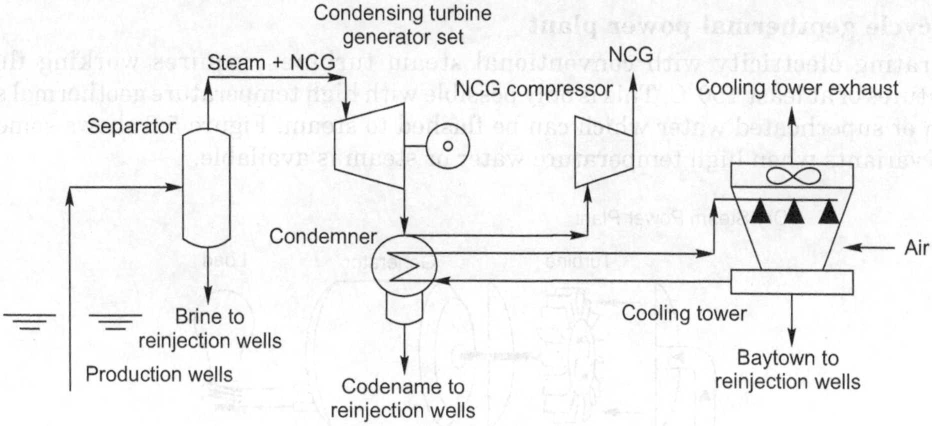

Fig. 5.7. Geothermal Power Plant

Main parts of Geothermal power plant

1. Production Well

   Hot GT water: Hot water is mined through production bore.

   Cooled GT water: Cooled GT water is being crowd into (re-injected) back to earth crust through re-injection bore. Re-injection avoids penetration of GT water containing huge volume of mineral and various harmful gases to surrounding environment. Bores mouths are generally closed by each other bottom terminals are away from each other to avoid cooling hot underground pool by cooled water. To reach the configuration oblique alternatively angulations bores are used.
2. Vaporizer: At vaporizer heat of GT water is used on transpiration of organic liquid with low boiling point. Aroused gas of high pressure is controlled to turbine.
3. Circulating Pump: Circulating pump drive condensate to vaporizer, where whole energy transformation round encore.
4. Expansion Turbine: It is driven by compressive energy of organic liquid vapour case of high pressure GT water, which becomes to be wet stream in expander steam turbine, is used to drive Generator.
5. Generator: This transforms Kinetic energy of turbine to electric. Used alternating current generators are mainly synchronous, alternatively asynchronous, synchronous generator are apt for electrically separated system, while asynchronous are assigned for parallel running with grid system.
6. Condenser: This improves efficiency of the cycle by cooling and liquefying Steam exiting steam turbine. Aroused hot liquid is swabbed back to vaporizer.
7. Transformer.

Geothermal Energy Conversion Systems

The thermal energy available from geothermal sources is harvested and used in various forms and equipment for utility scale geothermal power generation is available in capacities from as low as 100 kW .

1. Single Cycle Geothermal Power Plants
2. Binary Cycle Power Plants

Single cycle geothermal power plant

Generating electricity with conventional steam turbines requires working fluids at temperatures of at least 150°C. This is only possible with high temperature geothermal sources of steam or superheated water which can be flashed to steam. Figure 5.8 shows some of the possible variants when high temperature water or steam is available.

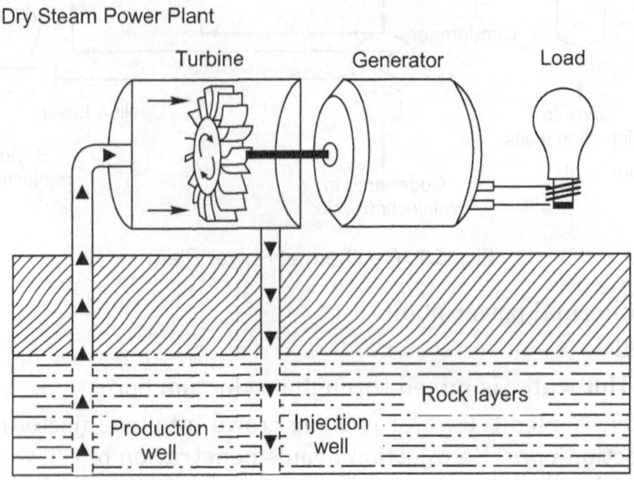

Fig. 5.8. Geothermal Power Generation – Dry/Flash Steam Systems

If steam is available directly from an aquifer it can also be used directly; after filtering, to drive a turbine. Otherwise steam can be generated from superheated water ejected by the aquifer. Turbines are available with either back -pressure (atmospheric) or condensing exhausts.

Condensing systems are more efficient but atmospheric exhaust turbine systems are simpler and cheaper. In a condensing system the condensate is usually re-injected into the ground, not necessarily into the aquifer. In an atmospheric system exhaust steam may be released to the atmosphere or used for other heating applications.

(*i*) *Dry Steam Systems:* Early geothermal systems made direct use of hot dry steam generated in hot aquifers and erupting from geysers with temperatures typically above 235°C, piped directly to steam turbines.

This is by far the cheapest method of generating electrical energy from geothermal resources. The largest plants have generating capacities of over 1000 MW. Unfortunately there are few suitable sources in the world and dry steam plants (Figure 5.15) are rare.

(*ii*) *Flash Steam Systems:* Although there are not many natural sources of hot dry steam, superheated water at high temperature and pressure is available from many more hot water aquifers as well as hot rock systems. The boiling point fluid increases as its pressure is increased and superheated water is liquid water under pressure at a temperature higher than the normal boiling point of 100°C.

When the pressure is reduced the water flashes to steam. Superheated water pumped from the ground at temperatures of 175°C or more can be flashed to steam in a separator or flash tank to drive a turbine directly. Surplus water from the flash plant is re-injected into the ground. The most common geothermal electricity generating plants are currently flash generators but most newly planned installations are expected to use binary systems.

Binary cycle power plants

In many geothermal and hydrothermal plants the water temperature and pressure are too low for driving a steam turbine efficiently if at all. Dual cycle "binary plants" Figure 5.9 have been developed to make more efficient use of lower temperature water in the range from 100°C to 175°C.

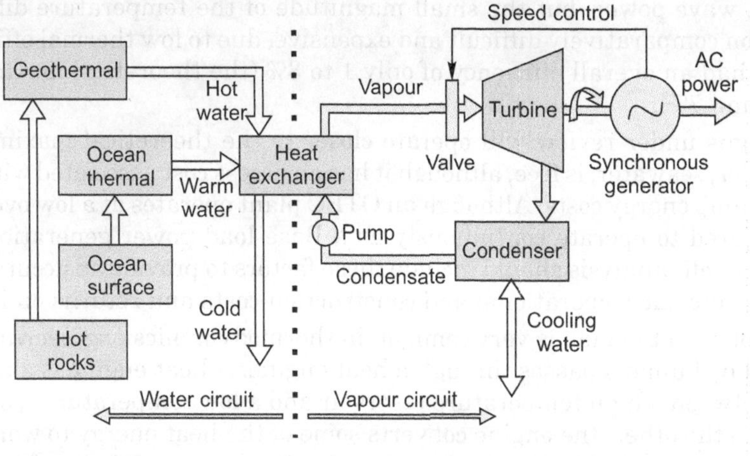

Fig. 5.9. Geothermal Power Generation—Binary System

In a binary plant the hot water circuit passing through the thermal. Source is separated Tom the closed loop working fluid circuit used in the turbine by a heat exchanger.

The hot water gives up its heat in the heat exchanger to a working fluid with a low boiling point and high vapour pressure at low temperatures when compared to steam.

The working fluid is typically an organic compound such as ammonia, butane pentane or isopentane which circulates through the secondary side of the heat exchanger where it vaporises and the vapour is then used to rotate a turbine in a conventional rankine cycle electricity generating plant.

After the vaporised binary liquid has passed through, and given up its energy to, the turbine it is condensed and recycled for re-use through the heat exchanger.

After passing through the heat exchanger the water in the primary circuit is pumped back down the bore hole into the hot rock. Binary systems have the added advantage that contaminated water from the primary heat source (salt water from the ocean or polluted water from aquifers and hot rocks) never comes in contact with the turbine generator units.

5.5 OCEAN THERMAL ENERGY CONVERSION (OTEC)

Ocean thermal energy conversion (OTEC) is a method for generating electricity which uses the temperature difference that exists between deep and shallow waters to run a heat engine. As with any heat engine, the greatest efficiency and power is produced with the largest temperature difference. This temperature difference generally increases with decreasing latitude. *i.e.*, near the equator, in the tropics.

Historically, the main technical challenge of OTEC was to generate significant amounts of power, efficiently, from this very small temperature ratio. Changes in efficiency of heat exchange in modem designs allow performance approaching the theoretical maximum efficiency.

The Earth's oceans are continually heated by the sun and cover nearly 70% of the Earth's surface; this temperature difference contains a vast amount of solar energy which can potentially be harnessed for human use. If this extraction could be made cost effective on a large scale, it could provide a source of renewable energy needed to deal with energy shortages, and other energy problems.

The total energy available is one or two orders of magnitude higher than other ocean energy options such as wave power, but the small magnitude of the temperature difference makes energy extraction comparatively difficult and expensive, due to low thermal efficiency. Earlier OTEC systems had an overall efficiency of only 1 to 3% (the theoretical maximum efficiency lies between 6 and 7%).

Current designs under review will operate closer to the theoretical maximum efficiency. The energy carrier, seawater, is free, although it has an access cost associated with the pumping materials and pump energy costs. Although an OTEC plant operates at a low overall efficiency, it can be configured to operate continuously as a Base load power generation system. Any thorough Cost-benefit analysis should include these factors to provide an accurate assessment of performance, efficiency, operational and construction costs and returns on investment.

The concept of a heat engine is very common in thermodynamics engineering, and much of the energy used by humans passes through a heat engine. A heat engine is a thermodynamic device placed between a high temperature reservoir and a low temperature reservoir. As heat flows from one to the other, the engine converts some of the heat energy to work energy. This principle is used in steam turbines and internal combustion engines. While refrigerators reverse the direction of flow of both the heat and work energy. Rather than using heat energy from the burning of fuel. OTEC power draws on temperature differences caused by the sun's warming of the ocean surface.

The only heat cycle suitable for OTEC, is the Rankine cycle, using a low-pressure turbine. Systems may be either closed-cycle or open-cycle. Closed-cycle engines use working fluids that are typically thought of as refrigerants such as ammonia or R -134a. Open-cycle engines use the water heat source as the working fluid.

Principles of Operation

If it could become cost-competitive with conventional power technologies, OTEC could produce gigawatts of electrical power, and in conjunction with electrolysis, could produce enough hydrogen to completely replace all projected global fossil fuel consumption. Managing costs is still a huge challenge, however. All OTEC plants require an expensive, large diameter intake pipe, which is submerged a kilometer or more into the ocean's depths, to bring very cold water to the surface.

Depending on the location the OTEC plants can be classified as:

- Land based plant
- Shelf based plant
- Floating plant

Depending on the cycle used the OTEC plants can be classified as:

- Open cycle
- Closed cycle
- Hybrid cycle

This cold seawater is an integral part of each of the three types of OTEC systems: closed-cycle, open-cycle, and hybrid. To operate, the cold seawater must be brought to the surface. This can be accomplished through direct pumping. A second method is to desalinate the seawater near the sea floor; this lowers its density, which will cause it to "float" up through a pipe to the surface.

Closed-cycle

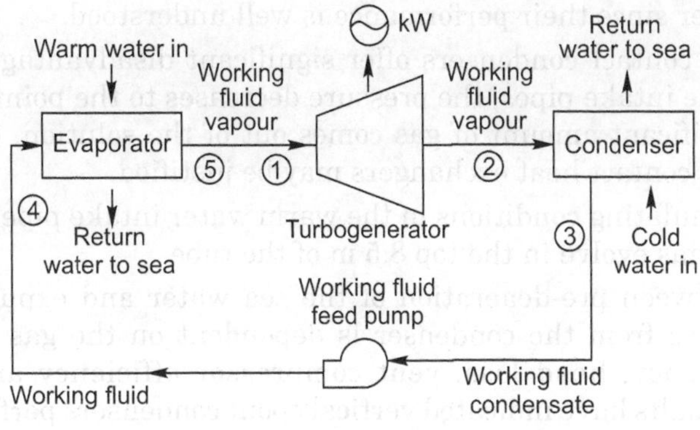

Fig. 5.10. Closed Cycle OTEC Plant

Closed-cycle systems (Figure 5.10) use fluid with a low boiling point, such as ammonia, to rotate a turbine to generate electricity. Warm surface seawater is pumped through a heat exchanger where the low-boiling-point fluid is vaporized. The expanding vapour turns the turbo-generator. Then, cold, deep seawater-pumped through a second heat exchanger-condenses the vapour back into a liquid, which is then recycled through the system.

Open-cycle

Open-cycle OTEC uses the tropical oceans' warm surface water to make electricity. When warm seawater is placed in a low-pressure container, it boils. The expanding steam drives a low-pressure turbine attached to an electrical generator. The steam, which has left its salt and contaminants behind in the low-pressure container, is pure fresh water. It is condensed back into a liquid by exposure to cold temperatures from deep-ocean water. This method has the advantage of producing desalinized fresh water, suitable for drinking water or irrigation.

Hybrid

A hybrid cycle combines the features of both the closed-cycle and open-cycle systems. In a hybrid OTEC system, warm seawater enters a vacuum chamber where it is flash-evaporated into steam, similar to the open-cycle evaporation process.

The steam vaporizes the ammonia working fluid of a closed-cycle loop on the other side of an ammonia vaporizer. The vaporized fluid then drives a turbine to produce electricity. The steam condenses within the heat exchanger and provides desalinated water.

The electricity produced by the system can be delivered to a utility grid or used to manufacture methanol, hydrogen, refined metals, ammonia, and similar products.

OTEC has important benefits other than power production. Some of them are:

- Air conditioning
- Chilled-soil agriculture
- Aquaculture
- Desalination
- Hydrogen production
- Mineral extraction

Disadvantages of OTEC

1. *Degradation of heat exchanger performance by dissolved gases:* A very important technical issue pertaining to the Claude cycle is the performance of direct contact heat exchangers operating at typical OTEC boundary conditions. Many early Claude cycle designs used a surface condenser since their performance is well understood.

   However, direct contact condensers offer significant disadvantages. As the warm sea water rises in the intake pipes, the pressure decreases to the point where gas begins to evolve. If a significant amount of gas comes out of the solution, designing a gas trap before the direct contact heat exchangers may be justified.

   Experiments simulating conditions in the warm water intake pipe indicated about 30% of the dissolved gas evolve in the top 8.5 m of the tube.

   The tradeoff between pre-deaeration of the sea water and expulsion of all the non-condensable gases from the condenser is dependent on the gas evolution dynamics, deaerator efficiency, head loss, vent compressor efficiency and parasitic power. Experimental results have indicated vertical spout condensers perform some 30% better than falling jet types.

2. *Degradation of heat exchanger performance by microbial fouling:* Because raw sea water must be passed through the heat exchanger care must be taken to maintain good thermal conductivity. Biofouling layers as thin as 25 to 50 μm can degrade heat exchanger performance by as much as 50%.

   A 1977 study in which mock heat exchangers were exposed to sea water for ten weeks concluded that although the level of microbial fouling was low, the thermal conductivity of the system was significantly impaired. The apparent discrepancy between the level of fouling and the heat transfer impairment is the result of a thin layer of water trapped by the microbial growth on the surface of the heat exchanger.

3. *Improper sealing:* The evaporator, turbine, and condenser operate in partial vacuum ranging from 3% to 1% atmospheric pressure. This poses a number of practical concerns.

   First, the system must be carefully sealed to prevent in-leakage of atmospheric air that can severely degrade or shut down operation.

   Second, the specific volume of low-pressure steam is very large compared to that of the pressurized working fluid used in the case of a closed cycle OTEC. This means components must have large flow areas to ensure steam velocities do not attain excessively high values.

4. *Parasitic power consumption by exhaust compressor:* An approach for reducing the exhaust compressor parasitic power loss is as follows. After most of the steam has been condensed by spout condensers, the non-condensable gas steam mixture is passed through a counter current region which increases the gas-steam reaction by a factor of five. The result is an 80% reduction in the exhaust pumping power requirements.

Sagar Shakthi: India's first 'power plant' on sea

Sagar Shakthi—the Ocean Thermal Energy Conversion (OTEC) Barge—a power plant, is the first of its kind in the world to generate electricity utilizing the temperature gradients between surface and deep-sea water. The barge is 68.5 m long, 16 m broad and 4 m deep, and houses the Rankine Cycle based power plant.

The barge has been jointly conceived and developed by the National Institute of Ocean Technology, Chennai, and Dempo Shipbuilding and Engineering Pvt. Ltd, Goa.

The principle used in OTEC barge is an innovated reverse process of refrigeration where the liquid ammonia carried at pressure in the system is vaporised in an evaporator using high temperature (28°C) surface sea water and the high energy vapour driving a sophisticated gas turbine generating electricity.

The vapour ammonia, after driving the generator, is then cooled back into liquid in a condenser using deep-sea water (7°C) and routed back into the system.

This is a technology demonstration pilot plant with provision for data collection on all aspects of the design and operation of the process, which would help developing commercial scale plants that can be deployed along the coast providing eco-friendly, renewable source of electricity, Simon affirms. This pilot plant can generate 1 MW of electricity.

Although some of the stages of the process have been tried out separately in laboratory and field conditions by developed countries, this is the first time an integrated full-sized plant encompassing the entire process has been built.

The barge has two moon pools with warm water and cold water boxes to take in the surface and deep-sea water for the plant. The deep-sea water is taken through a 1,200 m long HDPK pipeline suspended vertically in sea by means of solid ballast and mooring buoy. The OTEC barge has one of the deepest single point mooring systems in the world.

The project was built by Dempo Shipbuilding and Engineering at its shipyard at Bainguinim, Goa. The OTEC barge is presently anchored 40 nautical miles off the coast of Tuticorin Port.

5.6 BIOGAS GENERATION

Biogas is a good fuel. Have you thought how this is formed Biomass like animal excreta, vegetable wastes and weeds undergo decomposition in the absence of oxygen in a biogas plant and form a mixture of gases. This mixture is the biogas. Its main constituent is methane. This is used as a fuel for cooking and lighting.

Aerobic and Anaerobic Bioconversion Process

There are mainly three aerobic and anaerobic bioconversion process for the biomass energy applications: There are:

Bio-products: Converting biomass into chemicals for making products that typically are made from petroleum.

Biofuels: Converting biomass into liquid fuels for transportation.

Bio-power: Burning biomass directly, or converting it into a gaseous fuel or oil, to generate electricity.

Bio-products. Whatever products we can make from fossil fuels, we can make using biomass. These bio-products, or bio-based products, are not only made from renewable sources, they also often require less energy to produce than petroleum-based products.

Researchers have discovered that the process for making bio-fuels releasing the sugars that make up starch and cellulose in plants also can be used to make antifreeze, plastics, glues, artificial sweeteners, and gel for toothpaste.

Other important building blocks for bio-products include carbon monoxide and hydrogen. When biomass is heated with a small amount of oxygen present, these two gases are produced in abundance. Scientists call this mixture biosynthesis gas. Biosynthesis gas can be used to make plastics and acids, which can be used in making photographic films, textiles, and synthetic fabrics.

When biomass is heated in the absence of oxygen, it forms pyrolysis oil. A chemical called phenol can be extracted from pyrolysis oil. Phenol is used to make wood adhesives, molded plastic, and foam insulation.

Biofuels: Unlike other renewable energy sources, biomass can be converted directly into Liquid fuels, biofuels. For our transportation needs (cars, trucks, buses, airplanes, and trains). The two most common types of biofuels are ethanol and biodiesel.

Ethanol is an alcohol, the same found in beer and wine. It is made by fermenting any biomass high in carbohydrates (starches, sugars, or celluloses) through a process similar to brewing beer. Ethanol is mostly used as a fuel additive to cut down a vehicle's carbon monoxide and other smog-causing emissions. But flexible fuel vehicles, which run on mixtures of gasoline and up to 85% ethanol, are now available.

Biodiesel is made by combining alcohol (usually methanol) with vegetable oil, animal fat, or recycled cooking greases. It can be used as an additive to reduce vehicle emissions (typically 20%) or in its pure form as a renewable alternative fuel for diesel engines.

Other biofuels include methanol and reformulated gasoline components. Methanol, commonly called wood alcohol, is currently produced from natural gas, but could also be produced from biomass.

There are a number of ways to convert biomass to methanol, but the most likely approach is gasification. Gasification involves vaporizing the biomass at high temperatures, then removing impurities from the hot gas and passing it through a catalyst, which converts it into methanol.

Most reformulated gasoline components produced from biomass are pollution reducing fuel additives, such as methyl tertiary butyl ether (MTBE) and ethyl tertiary butyl ether (ETBE).

Bio-power: Bio-power, or biomass power, is the use of biomass to generate electricity. There are six major types of bio-power systems: direct fired, cofiring, gasification, anaerobic digestion, pyrolysis, and small, modular.

Most of the bio-power plants in the world use direct fired systems. They burn bioenergy feedstock's directly to produce steam. This steam is usually captured by a turbine, and a generator then converts it into electricity. In some industries, the steam from the power plant is also used for manufacturing processes or to heat buildings. These are known as combined heat and power facilities. For instance, wood waste is often used to produce both electricity and steam at paper mills.

Many coal fired power plants can use cofiring systems to significantly reduce emissions, especially sulphur dioxide emissions. Coal firing involves using bioenergy feedstocks as a supplementary energy source in high efficiency boilers.

Gasification systems use high temperatures and an oxygen starved environment to convert biomass into a gas (a mixture of hydrogen, carbon monoxide, and methane). The gas fuels what's called a gas turbine, which is very much like a jet engine, only it turns an electric generator instead of propelling a jet.

The decay of biomass produces a gas methane that can be used as an energy source. In landfills, wells can be drilled to release the methane from the decaying organic matter. Then pipes from each well carry the gas to a central point where it is filtered and cleaned before burning. Methane also can be produced from biomass through a process called anaerobic digestion. Anaerobic digestion involves using bacteria to decompose organic matter in the absence of oxygen.

Methane can be used as an energy source in many ways. Most facilities burn it in a boiler to produce steam for electricity generation or for industrial processes. Two new ways include the use of microturbines and fuel cells. Microturbines have outputs of 25 to 500 kilowatts. About the size of a refrigerator, they can be used where there are space limitations for power production. Methane can also be used as the "fuel" in a fuel cell. Fuel cells work much like batteries but never need recharging, producing electricity as long as there's fuel.

In addition to gas, liquid fuels can be produced from biomass through a process called pyrolysis. Pyrolysis occurs when biomass is heated in the absence of oxygen. The biomass then turns into a liquid called pyrolysis oil, which can be burned like petroleum to generate electricity. A bio-power system that uses pyrolysis oil is being commercialized.

Several bio-power technologies can be used in small, modular systems. A small, modular system generates electricity at a capacity of 5 megawatts or less. This system is designed for use at the small town level or even at the consumer level. For example, some farmers use the waste from their livestock to provide their farms with electricity. Not only do these systems provide renewable energy, they also help farmers and ranchers meet environmental regulations.

Small, modular systems also have potential as distributed energy resources. Distributed energy resources refer to a variety of small, modular power generating technologies that can be combined to improve the operation of the electricity delivery system.

Raw Materials

All types of organic wastes which can form slurry are suitable for producing biogas by the process of anaerobic digestion in a biogas plant. Wood and sugar biogases are difficult and time consuming with this process and incineration may be preferred. The choice of raw material (in feed) is based on availability of the waste. The biogas plant is (Figure 5.11) designed to suit particular type of in feed.

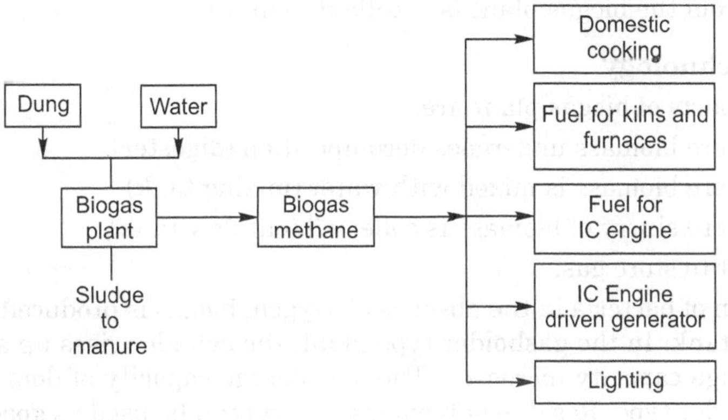

Fig. 5.11. Energy Route of Biogas (Gobar Gas)

Biogas production taken different time period depending upon raw material; temperature; process adopted, etc.

The biomass used as a raw material can be classified into the following categories.

Waste	*Cultivated and Harvested*
Agricultural wastes	Agricultural energy crops
Rural animal wastes	Aquatic crops
Poultry waste	
Butchary waste	
Urban waste (garbage)	Forest crops
Aquatic wastes	
Forest wastes	
Coconut husk waste	
Industrial wastes	

Others are poultry waste, piggery waste, sheep, goat, cow, horse dung, Slaughter house waste, coconut shell, husk, waste garbage, fruit skins and leftovers.

The waste is generated periodically and can be converted into useful biogas. The problem of waste disposal is solved as the sludge is used as manure.

Waste ——— Biogas plant ——— Biogas

|

Sluge manure

The cultivated or harvested biomass is specially grown on land or in sea/lake for obtaining raw materials for biogas production.

Properties of Biogas

Main properties of biogas are:

1. Comparatively simple and can be produced easily.
2. Burns without smoke and without leaving ash as residues.
3. Household wastes and bio-wastes can be disposed of usefully and in a healthy manner.
4. Reduces the use of wood and to a certain extent prevents deforestation.
5. The slurry from the biogas plant is excellent manure.

Biogas Plant Technology

The important parts of biogas plant are:

1. The tank where biomass undergoes decomposition (digester)
2. The tank where biomass is mixed with water (mixing tank)
3. The tank where slurry of biomass is collected (out flow tank)
4. Arrangement to store gas.

Due to the action of bacteria in the absence of oxygen, biogas is produced in the plant. This is collected in the tank. In the gasholder type plant, the cylinder rises up as the gas fills the tank and the storage capacity increases. The gas storage capacity of dome type will be less than that of gasholder type. Residue of biomass (slurry) can be used as good manure.

Biogas plants are built in several sizes, small (0.5 m^3/day) to very large 2500 m^3/day). Accordingly, the configurations are simpler to complex.

Biogas plants are classified into following main types:

—Continuous type or batch type.

—Drum type and dome type.

There are various configurations within these types.

Continuous Type

Continuous type biogas plant delivers the biogas continuously and is fed with the biomass regularly. Continuous type biogas plant is of two types.

(A) Single Stage Continuous Type Biogas Plant

In such a plant Phase-I (acid formation) and Phase-II (methanation) are carried out in the same chamber without barrier. Such plants are simple, economical, easy to operate and control. These plants are generally preferred for small and medium size biogas plants. Single stage plants have lesser rate of gas production than the two stage plant.

(B) Two State Continuous Type Biogas Plant

In such a plant the Phase-I (acid formation) and Phase-II (methane formation) take place in separate chambers. The plant produces more biogas in the given time than the single stage plant. However, the process is complex and the plant is costlier, difficult to operate and maintain. Two stage plant is preferred for larger biogas plant systems.

Batch Type Biogas Plant

The infeed biomass is fed in batches with large time interval between two consecutive batches. One batch of biomass infeed is given sufficient **retention** time in the digester (30 to 50 days). After completion of the digestion, the residue is emptied and the fresh charge is fed. The fresh biomass charge may be subjected to aeration or nitrogenation after feeding and then the digester covers are closed for the digestion process. Thereafter, the Biogas is derived from the digester after 10 to 15 days. Fermentation continues for 30 to 50 days.

Salient Features:

1. Batch type biogas plant delivers gas intermittently and discontinuously.
2. Batch type biogas plant may have several digesters (reacters) which are fed in a sequential manner and discharged in a sequential manner to obtain the output biogas continuously.
3. Batch type biogas plants have longer digestion time and are therefore more suitable for materials which are difficult for anaerobic digestion (*e.g.* harder, fibrous biomass).
4. Batch type biogas plant needs initial seeding to start the anaerobic fermentation.
5. Batch type biogas plant needs larger volume of the digester to accommodate large volume of the batch. Hence initial cost is higher.
6. Operation and maintenance is relatively more complex. Batch type biomass plants need well organized and planned feeding. Such plants are preferred by European farmers. Such plants are not yet popular in India.

Fixed Dome Type Digester

In the fixed dome type digester biogas plant, the digester and gas-collector (gas dome) are enclosed in the same chamber. This type of construction is suitable for batch type biogas plant. The digester is conveniently built at or below ground level in comparatively cooler zone. The construction of the digester is with locally available materials like, bricks, teracota. The pressure inside the digester increases as the biogas is liberated. The biogas gets collected in the upper portion of the digester in a dome shaped cavity. The outlet pipe is provided at the tope of the fixed dome. Alternatively the gas collector (gas holder) is a separately installed chamber. The digester tank and gas collector chamber are separated by a water seal tank.

The arrangement of a separate gas collector is preferred as the tapping of gas from the gas holder does not affect the pressure and the digestion process in the main digester. The water seal tank prevents the return of the gas from the gas collector to the digester chamber.

An additional displacement chamber may be provided for providing space to the displacement slurry in the digester due to gas pressure in the upper dome of the fixed type digester. The fixed dome type digester can be fed on daily basis with small quantities of the slurry. The excess slurry in the digester gets accommodated in the displacement chamber. The level of the slurry in the main digester and the displacement collector can vary in accordance with the pressure and volume of the biogas in the fixed type of dome. The pressure in the fixed dome and the displacement gas collector are almost the same as they are connected by the outlet from the main digester.

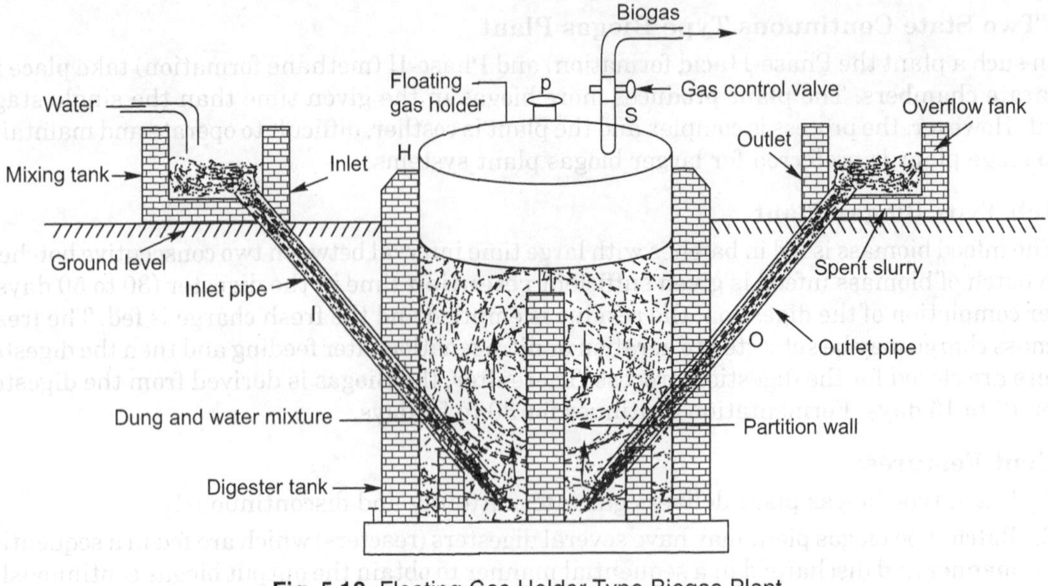

Fig. 5.12. Floating Gas Holder Type Biogas Plant

Floating Gas Holder Type. In this design a dome made floats above the slurry in the disaster. In the Figure 5.12, the disaster tank is of cylindrical masonry construction. The floating dome is of fabricated steel construction. The dome guide shaft provides the axial guide to the floating dome. As the gas is collected in it. The sliding bearing provides smooth sliding surface and guide to the floating dome. The gas generated in the slurry gets collected in the dome and the dome arises. The water seal tank provides separation between the gas in the dome and the outlet gas.

5.7 FUEL CELLS

In a fuel cell, electrical energy is generated directly through the electrochemical reaction of oxidant (oxygen from air) and fuels (such as natural gas, methanol, or pure hydrogen) at two electrodes separated by an electrolyte. When pure hydrogen is used, the only products of this process are heat, electricity and water. Unlike a battery, fuel cells do not store energy. The energy conversion is achieved without making use of materials that are stored within, or constitute an integral part of the fuel cell structure. It should be noted that fuel cells convert chemical energy directly into electricity without an intermediate combustion process.

The operation of a fuel cell requires a fuel electrode (*Anode*), oxidant electrode (*Cathode*), electrically-insulating and ionic conductor (*Electrolyte*), and external and electrically conducting circuit. In general, fuel cells can be classified according to the type of ionic conductor (electrolyte) they use and the temperature range at which they operate. Figure 5.13 provides a brief summary of various types of fuel cells.

Several types of fuel cells are currently under development. Alkali fuel cells (AFC), being used for a long time by NASA on space missions, use alkaline potassium hydroxide as the electrolyte. In proton exchange membrane fuel cell (PEMFC) and phosphoric acid fuel cell (PAFC), hydrogen fuel dissociates into free electrons and protons (positive hydrogen ions). The hydrogen protons migrate through the electrolyte to the cathode. At the cathode, oxygen from air, electrons from the external circuit and protons combine to form pure water and heat. All these three types are low temperature fuel cells.

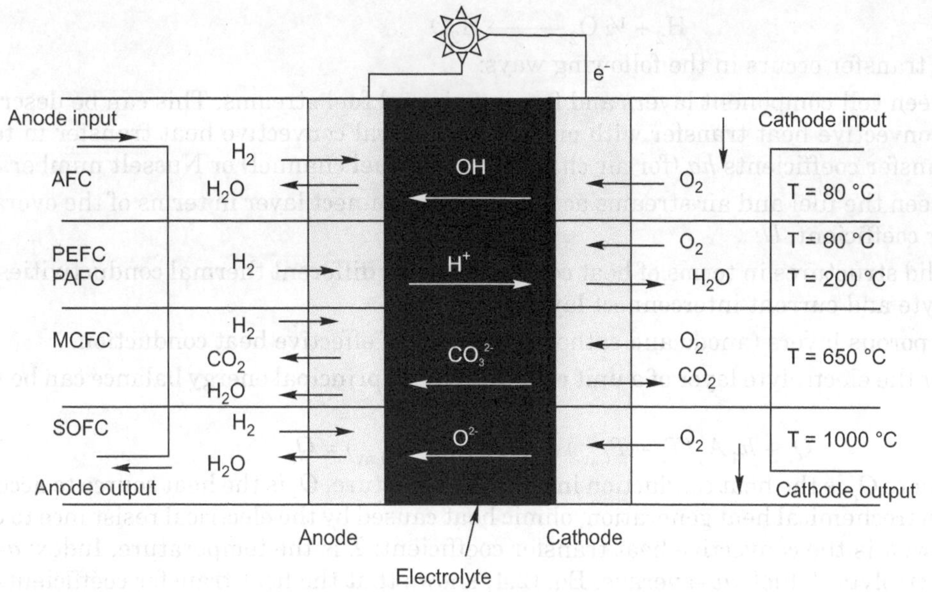

Fig. 5.13. Summary of Various Types of Fuel Cells

High temperature fuel cells, such as solid oxide fuel cells (SOFCs) and molten carbonate ones (MCFC) are of particular interest, because operation at high temperatures allows usage of natural gas as a fuel, and hybrid concepts involving a combination of a fuel cell and a gas turbine. The overall system efficiency can be significantly increased. Operation at temperature about 1000 *DC* (conventional, electrolyte-supported design), 700 *DC* (intermedium temperature, anode-supported design) and pressures greater than one atmosphere leads to solid oxide fuel cells (SOFCs) as the obvious choice [1, 2].

In the anode duct, fuel (*e.g.*, *Hd* is supplied and air (O_2+ N_2) is introduced in the cathode duct, and these ducts are separated by the electrolyte/electrode assembly. Reactants are transported by diffusion and/or convection to the electrode/electrolyte (SOFC) or catalyst/ electrolyte (PEMFC) interfaces, where electrochemical reactions take place. An electrochemical oxidation reaction at the anode produces electrons that flow through the inter-collector (bipolar plate, for PEMFC) or -connector (for SOFC) to the external circuit, while the ions pass through the electrolyte to the opposing electrode. The electrons return from the external circuit to participate in the electrochemical reduction reaction at the cathode. In the electrochemical reaction process, part of the oxygen is consumed in the cathode duct, while the hydrogen is consumed in the anode duct. Heat and water *(H_2O)* are only the by-products during the process. The water generated is injected into the anode duct further along the duct in SOFCs, while in PEMFCs it enters into the cathode duct. The electrochemical reactions can be written as Eqs. (2.1 *a, b)* in SOFCs:

Cathode reaction: $\frac{1}{2} O_2 + 2e^- \longrightarrow O^{2-}$ ...(2.1*a*)

Anode reaction: $H_2 + O^{2-} \longrightarrow H_2O + 2e^-$ ...(2.1*b*)

Equations (2.2*a*–*b*) describe the reactions in PEMFCs:

Cathode reaction: $O_2 + 4e^- + 4H^+ \longrightarrow 2H_2O$ ...(2.2*a*)

Anode reaction: $2H_2 + O_2 \longrightarrow 2H_2O$ ...(2.2*b*)

The overall reaction is as follows:

$$H_2 + ½ O_2 \longrightarrow H_2O \qquad ...(2.3)$$

Heat transfer occurs in the following ways:

Between cell component layers and flowing air and fuel streams. This can be described by forced convective heat transfer with or without natural convective heat transfer in terms of heat transfer coefficients *ha* (for air channel), *h*,(for fuel channel) or Nusselt numbers, *Nu;*

Between the fuel and air streams across the interconnect layer in terms of the overall heat transfer coefficient, *U;*

In solid structures in terms of heat conduction with different thermal conductivities, λ_1 (*i* = electrolyte and current interconnect layer);

- In porous layers (anode and cathode) in terms of effective heat conduction.

  For the electrolyte layer of a unit cell, the overall principal energy balance can be written as:

$$Q_c + h_1 A_1 (T_e - T_{f.av}) + h_a A_a (T_e - T_{a.av}) = Q_s \qquad ...(2.4)$$

Where Q_c is the heat conduction in the solid structure, Q_s is the heat source to account for electrochemical heat generation, ohmic heat caused by the electrical resistance to current flow; *h* is the convective heat transfer coefficient; *T* is the temperature. Index: *a*–air, *e*–electrolyte, *f*–fuel, *av*–average. Eq. (2.4) shows that the heat transfer coefficients in the fuel and oxidant ducts are important.

Due to the flow resistance in the fuel cells, pressure drop (ΔP) along the ducts and in manifolds can cause non-uniform distribution of flow rate. However, the temperature is always non-uniform even when one has equal mass flow in the ducts. This is caused by the heat transfer and phase change (in PEMFCs), and the available L_1 ΔT changing accordingly. Furthermore, the output of electrical energy would differ in terms of voltage potential and even gas re-circulation. At some severe conditions, the lack of gas in some channels can cause the irreversible damage to the fuel cell components. Pressure drop is depending on the channel and manifold structures, flow streams, etc.

5.8 MHD POWER GENERATION

Magneto hydrodynamic power generation provides a way of generating electricity directly from a fast moving stream of ionised gases without the need for any moving mechanical parts - no turbines and no rotary generators. Several MHD projects were initiated in the 1960s but overcoming the technical challenges of making a practical system proved very expensive. Interest consequently waned in favour of nuclear power which since that time has seemed a more attractive option.

MHD power generation has also been studied as a method for extracting electrical power from nuclear reactors and also from more conventional fuel combustion systems.

Working Principle

The MHD generator can be considered to be a fluid dynamo. This is similar to a mechanical dynamo in which the motion of a metal conductor through a magnetic field creates a current in the conductor except that in the MHD generator the metal conductor is replaced by a conducting gas plasma. The principle of magnetohydrodynamic power generation is illustrated in Fig. 5.14.

When a conductor moves through a magnetic field it creates an electrical field perpendicular to the magnetic field and the direction of movement of the conductor. This is the principle, discovered by Michael **Faraday**, behind the conventional rotary electricity generator. Dutch physicist Antoon **Lorentz** provided the mathematical theory to quantify its effects.

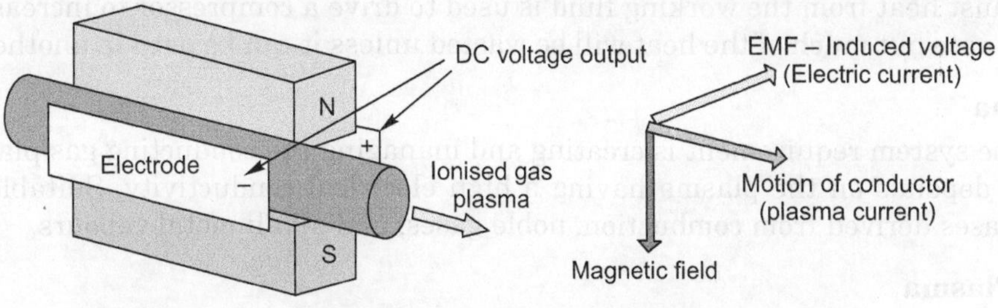

Fig. 5.14. Magnetohydrodynamic Power Generation (Principle)

The flow (motion) of the conducting plasma through a magnetic field causes a voltage to be generated (and an associated current to flow) across the plasma, perpendicular to both the plasma flow and the magnetic field according to Fleming's Right Hand Rule.

Lorentz Law describing the effects of a charged particle moving in a constant magnetic field can be stated as:

$$F = QvB$$

where

F is the force acting on the charged particle

Q is charge of particle

v is velocity of particle

B is magnetic field

The MHD System

The MHD generator needs a high temperature gas source, which could be the coolant from a nuclear reactor or more likely high temperature combustion gases generated by burning fossil fuels, including coal, in a combustion chamber. The Fig. 5.15 shows possible system components.

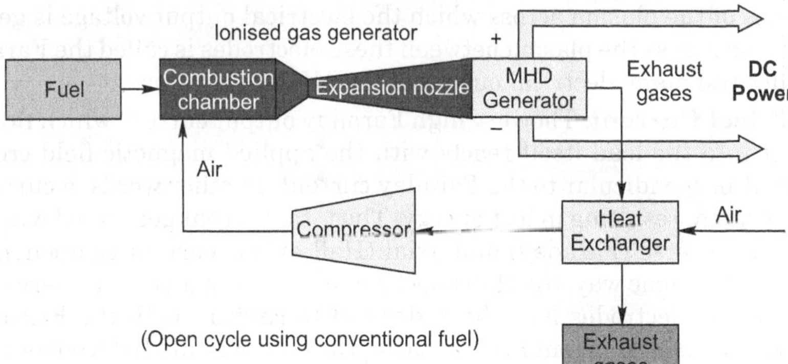

Fig. 5.15. Magnetohydrodynamic (MHD) Electricity Generation

The expansion nozzle reduces the gas pressure and consequently increases the plasma speed (Bernoulli's Law) through the generator duct to increase the power output (See Power below). Unfortunately, at the same time, the pressure drop causes the plasma temperature to fall (Gay-Lussac's Law) which also increases the plasma resistance, so a compromise between Bernoulli and Gay-Lussac must be found.

The exhaust heat from the working fluid is used to drive a compressor to increase the fuel combustion rate but much of the heat will be wasted unless it can be used in another process.

The Plasma

The prime system requirement is creating and managing the conducting gas plasma since the system depends on the plasma having a high electrical conductivity. Suitable working fluids are gases derived from combustion, noble gases, and alkali metal vapours.

The Gas Plasma

To achieve high conductivity, the gas must be ionised, detaching the electrons from the atoms or molecules leaving positively charged ions of the gas. The plasma flows through the magnetic field at high speed, in some designs, more than the speed of sound, the flow of the charged particles providing the necessary moving electrical conductor.

Methods of Ionizing the Gas

Various methods for ionizing the gas are available, all of which depend on imparting sufficient energy to the gas. It may be accomplished by heating or irradiating the gas with *X*-rays or Gamma rays. It has also been proposed to use the coolant gases such as helium and carbon dioxide employed in some nuclear reactors as the plasma fuel for direct MHD electricity generation rather than extracting the heat energy of the gas through heat exchangers to raise steam to drive turbine generators. Seed materials such as Potassium carbonate or Cesium are often added in small amounts, typically about 1% of the total mass flow to increase the ionization and improve the conductivity, particularly of combustion gas plasmas.

Containment

Since the plasma temperature is typically over 1000°C, the duct containing the plasma must be constructed from non-conducting materials capable of withstanding these high temperatures. The electrodes must of course be conducting as well as heat resistant.

- **The Faraday Current:** A powerful electromagnet provides the magnetic field through which the plasma flows and perpendicular to this field are installed the two electrodes on opposite sides of the plasma across which the electrical output voltage is generated. The current flowing across the plasma between these electrodes is called the Faraday current. This provides the main electrical output of the MHD generator.
- **The Hall Effect Current:** The very high Faraday output current which flows across the plasma duct into the load itself reacts with the applied magnetic field creating a Hall Effect current perpendicular to the Faraday current, in other words, a current along the axis of the plasma, resulting in lost energy. The total current generated will be the vector sum of the transverse (Faraday) and axial (Hall effect) current components. Unless it can be captured in some way, the Hall effect current will constitute an energy loss. Various configurations of electrodes have been devised to capture both the Faraday and Hall effect components of the current in order to improve the overall MHD conversion efficiency.

  One such method is to split the electrode pair into a series of segments physically side by side (parallel) but insulated from each other, with the segmented electrode pairs connected in series to achieve a higher voltage but with a lower current. Instead of the electrodes being directly opposite each other, perpendicular to the plasma stream, they are skewed at a slight angle from perpendicular to be in line with the vector sum of the Faraday and Hall effect currents, as shown in the diagram below, thus allowing the maximum energy to be extracted from the plasma. MHD Current Flow with Segmented Electrodes is shown in Figure 5.16.

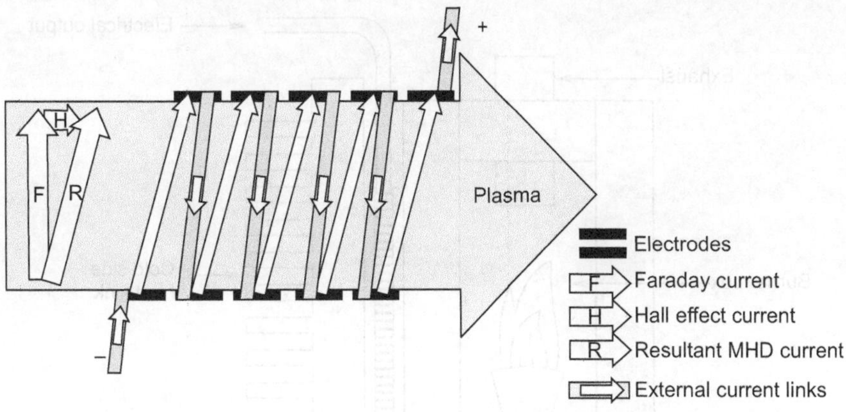

Fig. 5.16. MHD Current Flow with Segmented Electrodes

- **Power Output:** The output power is proportional to the cross-sectional area and the flow rate of the ionised plasma. The conductive substance is also cooled and slowed in this process. MHD generators typically reduce the temperature of the conductive substance from plasma temperatures to just over 1000°C. An MHD generator produces a direct current output which needs an expensive high power inverter to convert the output into alternating current for connection to the grid.
- **Efficiency:** Typical efficiencies of MHD generators are around 10 to 20 per cent mainly due to the heat lost through the high temperature exhaust.

  This limits the MHD's potential applications as a stand alone device but they were originally designed to be used in combination with other energy converters in hybrid applications where the output gases (flames) are used as the energy source to raise steam in a steam turbine plant. Total plant efficiencies of 65% could be possible in such arrangements.
- **Experience:** Demonstration plants with capacities of 50 MW or more have been built in several countries but MHD generators are expensive. Typical use could be in peak shaving applications but they are less efficient than combined-cycle gas turbines which means there are very few installations and MHD is currently not considered for mainstream commercial power generation.

5.9 TYPICAL CONFIGURATION OF THERMOELECTRIC POWER GENERATION SYSTEM

There are four basic components: a heat source, a thermoelectric power generation module (*i.e.*, a thermoelectric generator also known as a Seebeck device), a 'cold-side' heat sink, and the electrical load. The system may also include a voltage regulation circuit, or a fan for the heat sink.

The illustration in above Figure 5.17 shows one example. In this case we have a burner box with a propane fuel source. It is shown with the burner box open on one end, but in reality, it would be enclosed.

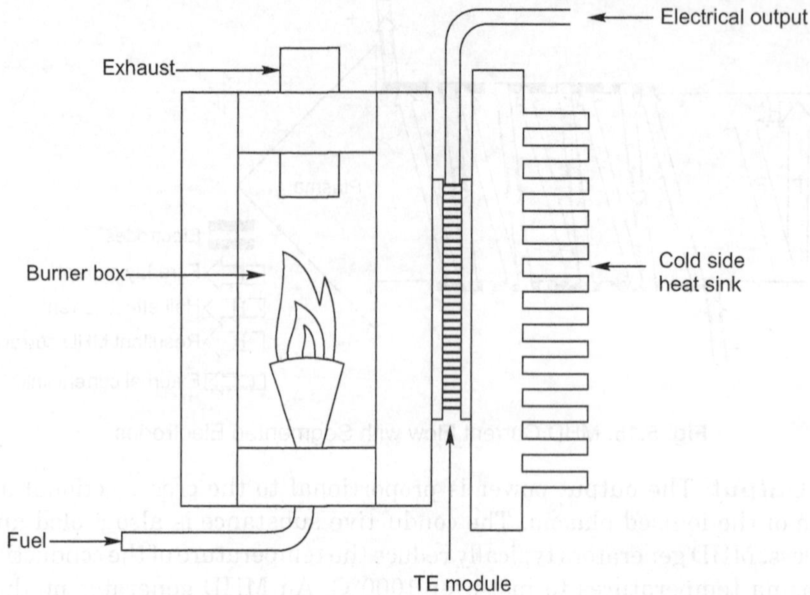

Fig. 5.17. Configuration of Thermoelectric Power Generation System

The thermoelectric module is then sandwiched between the heat source and the cold-side sink. While this example shows only a single thermoelectric power generation module, in reality. Several modules might be deployed in whatever series/parallel electrical arrangement best served the load.

It would be virtually impossible to get an adequate *DT* without some type of heat sink. However, you can sometimes reduce the size requirement for the sink (*i.e.*, film surface area) if you can find a way to insure good air flow.

First and foremost you want to prevent the hot side temperature of the thermoelectric device from exceeding the melting temperature of the solder employed to secure the semiconductor pellets to the copper tabs.

It is recommended that the temperature be kept below 2000 C. Toward this end, it is a good idea to use some type of 'heat spreader' to prevent hot spots at the hot side module interlace. Usually this means employing a relatively thick casting or extrusion between the heat source and the module.

On the mechanical side-especially when using multiple devices you need to find a means of applying compression between the hot and cold sides, which will apply even pressure across the thermoelectric modules and most importantly, prevent the hot side interface from bowing. If there is too great an expanse between compression points, the hot side interface can distort to the point where some modules are crushed or the thermal interface is compromise.

T_{Hot} is the temperature at the mounting surface of the thermoelectric module, which comes in contact with the heat source (*i.e.*, the hot side of the system).

T_{Cold} is the temperature at the mounting surface of the module, which comes in contact with the cold side heat sink.

No-load voltage (VNL) is the voltage output of the thermoelectric power generation system when no electrical load is connected.

Load voltage (VL) is the voltage output of the TEG system when all electrical load is connected.

Internal resistance (R_{Int}) is the electrical resistance of the thermoelectric power generation module (or module array).

Power conversion efficiency is the ratio of power output to power input expressed as a percentage. In this case, power output would be the wattage dissipated in the electrical load and power input would be the rate of energy use *e.g.*, watts to treate the necessary DT.

Electrical efficiency is the ratio of electrical power dissipated in the load to the total amount of power generated (including the dissipation in the internal resistance).

Solar Thermal Systems

Solar energy in one form or another is the source of nearly all energy on the earth. Humans, like all other animals and plants, rely on the sun for warmth and food. However, people also harness the sun's energy in many other different ways. For example, fossil fuels, plant matter from a past geological age, is used for transportation and electricity generation and is essentially just stored solar energy from millions of years ago. Similarly, biomass converts the sun's energy into a fuel, which can then be used for heat, transport or electricity. Wind energy, used for hundreds of years to provide mechanical energy or for transportation, uses air currents that are created by solar heated air and the rotation of the earth. Today wind turbines convert wind power into electricity as well as its traditional uses. Even hydroelectricity is derived from the sun. Hydropower depends on the evaporation of water by the sun, and its subsequent return to the Earth as rain to provide water in dams. Photovoltaics (often abbreviated as PV) are a simple and elegant method of harnessing the sun's energy. PV devices (solar cells) are unique in that they directly convert the incident solar radiation into electricity, with no noise, pollution or moving parts, making them robust, reliable and long lasting. Solar cells are based on the same principles and materials behind the communications and computer revolutions, and this CDROM covers the operation, use and applications of photovoltaic devices and systems.

Solar Cell

A solar cell is an electronic device which directly converts sunlight into electricity. Light shining on the solar cell produces both a current and a voltage to generate electric power. This process requires firstly, a material in which the absorption of light raises an electron to a higher energy state, and secondly, the movement of this higher energy electron from the solar cell into an external circuit. The electron then dissipates its energy in the external circuit and returns to the solar cell. A variety of materials and processes can potentially satisfy the requirements for photovoltaic energy conversion, but in practice nearly all photovoltaic energy conversion uses semiconductor materials in the form of a *p-n* junction.

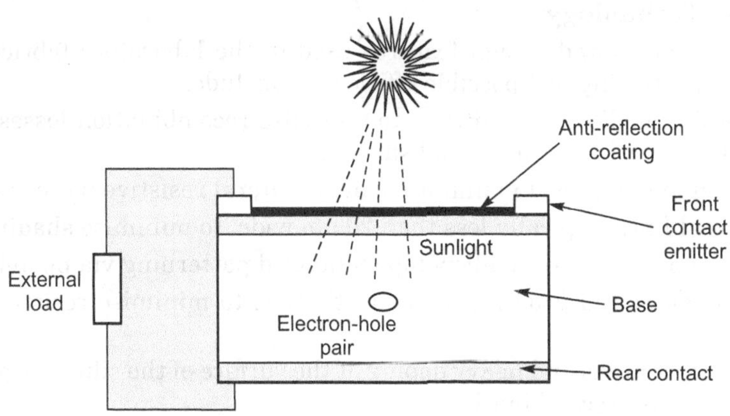

Fig. 5.18. (a) Cross-section of a solar cell

The basic steps in the operation of a solar cell are:

- the generation of light-generated carriers;
- the collection of the light-generated carries to generate a current;
- the generation of a large voltage across the solar cell; and
- the dissipation of power in the load and in parasitic resistances.

The generation of current in a solar cell, known as the "light-generated current", involves two key processes. The first process is the absorption of incident photons to create electron-hole pairs. Electron-hole pairs will be generated in the solar cell provided that the incident photon has an energy greater than that of the band gap. However, electrons (in the *p*-type material), and holes (in the *n*-type material) are meta-stable and will only exist, on average, for a length of time equal to the minority carrier lifetime before they recombine. If the carrier recombines, then the light-generated electron-hole pair is lost and no current or power can be generated.

A second process, the collection of these carriers by the *p-n* junction, prevents this recombination by using a *p-n* junction to spatially separate the electron and the hole. The carriers are separated by the action of the electric field existing at the *p-n* junction. If the light-generated minority carrier reaches the *p-n* junction, it is swept across the junction by the electric field at the junction, where it is now a majority carrier. If the emitter and base of the solar cell are connected together (*i.e.*, if the solar cell is short-circuited), the light-generated carriers flow through the external circuit.

The collection of light-generated carriers does not by itself give rise to power generation. In order to generate power, a voltage must be generated as well as a current. Voltage is generated in a solar cell by a process known as the "photovoltaic effect". The collection of light-generated carriers by the *p-n* junction causes a movement of electrons to the *n*-type side and holes to the *p*-type side of the junction. Under short circuit conditions, there is no build up of charge, as the carriers exit the device as light-generated current.

However, if the light-generated carriers are prevented from leaving the solar cell, then the collection of light-generated carriers causes an increase in the number of electrons on the *n*-type side of the *p-n* junction and a similar increase in holes in the *p*-type material. This separation of charge creates an electric field at the junction which is in opposition to that already existing at the junction, thereby reducing the net electric field. Since the electric field represents a barrier to the flow of the forward bias diffusion current, the reduction of the electric field increases the diffusion current. A new equilibrium is reached in which a voltage exists across the *p-n* junction.

Cell Fabrication Technology

Some of the techniques and design features used in the laboratory fabrication of silicon solar cells, to produce the highest possible efficiencies include:

1. lightly phosphorus diffused emitters, to minimise recombination losses and avoid the existence of a "dead layer" at the cell surface;
2. closely spaced metal lines, to minimise emitter lateral resistive power losses;
3. very fine metal lines, typically less than 20 μm wide, to minimise shading losses;
4. polished or lapped surfaces to allow top metal grid patterning via photolithography;
5. small area devices and good metal conductivities, to minimise resistive losses in the metal grid;
6. low metal contact areas and heavy doping at the surface of the silicon beneath the metal contact to minimise recombination;

7. use of elaborate metallization schemes, such as titanium/palladium/silvcr, that give very low contact resistances;
8. good rear surface passivation, to reduce recombination;
9. Use of anti-reflection coatings, which can reduce surface reflection from 30% to well below 10%.

A number of cell designs exist which incorporate advanced laboratory features. Two approaches that have been used by niche markets such as solar cars are the PERL cells produced at University of New South Wales, and the rear-contact cells developed at Stanford University and Sun Power.

Solar thermal system (Fig. 5.18) converts sunlight into heat for various forms of end-use. One of three different conversion cycles may be used in a thermoelectric system. The Seebeck effect results when the junctions of a loop made of two different materials are at different temperatures such that an electromotive force (emf), and consequently, a current is produced in the loop. If T_h is the hot junction temperature, T_c the cold junction temperature, then dV_{12}, the open circuit voltage, is related to the temperature difference by the Seebeck coefficienct as represented by

$$\alpha_s = \lim \Delta V_{12} = dV_{12}$$
$$(T_{hc} - T_{c)}\, T_h - T_c\, dT$$

When a current is passed through a loop of two different materials, one junction becomes hot and the other junction becomes cold. This is the Peltier effect. The Peltier coefficient a P represents the ratio of heat change at the junction to the current flow as shown by

$$\alpha_p = \frac{\dot{Q}_{12}}{I}$$

When a current flows through a conductor in which a temperature gradient exists, heat is either liberated or absorbed, depending on the direction of current flow. This is known as the Thomson effect. The Thomson coefficient a T is the ratio of heat change per unit of current flow to the local temperature, as shown by

$$\alpha_T = \frac{\Delta Q / \Delta x}{I\,(\Delta T / \Delta x)}$$

These three effects are reversible and are interrelated.

Solar collectors make use of one of these effects. There are two types of collectors flat plate and focusing. Flat plate collectors, illustrated in Figure 5.18, do not use any device for concentrating the sun's rays. These collectors still function when clouds cut off direct sunlight. This advantage, along with their favorable cost, allows flat plate collectors to be used for low temperature heating of up to 100°C.

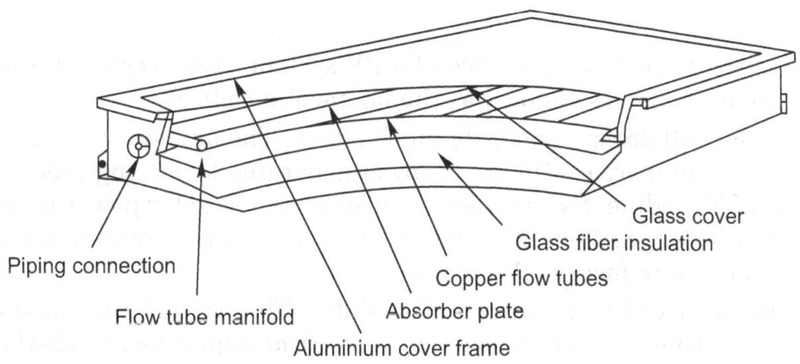

Fig. 5.18. (*b*) Solar Thermal Systems

Simpler flat plate collectors hold all the water that is to be heated, while more advanced ones heat only a little water at a time. The heated water is kept in a separate storage tank to cut down heat losses from the fluid. Common uses of flat plate collectors are for solar water heating, solar space heating and solar cooling, illustrated in Figure 5.19 (*a*), allow the use of much higher temperatures than available from the best flat plate collectors. A focusing collector comprises a concentrator and a receiver. The concentrator is the optical system that directs the sun's rays to the receiver. This could be a parabolic dish with a point focus or a parabolic trough with a linear focus.

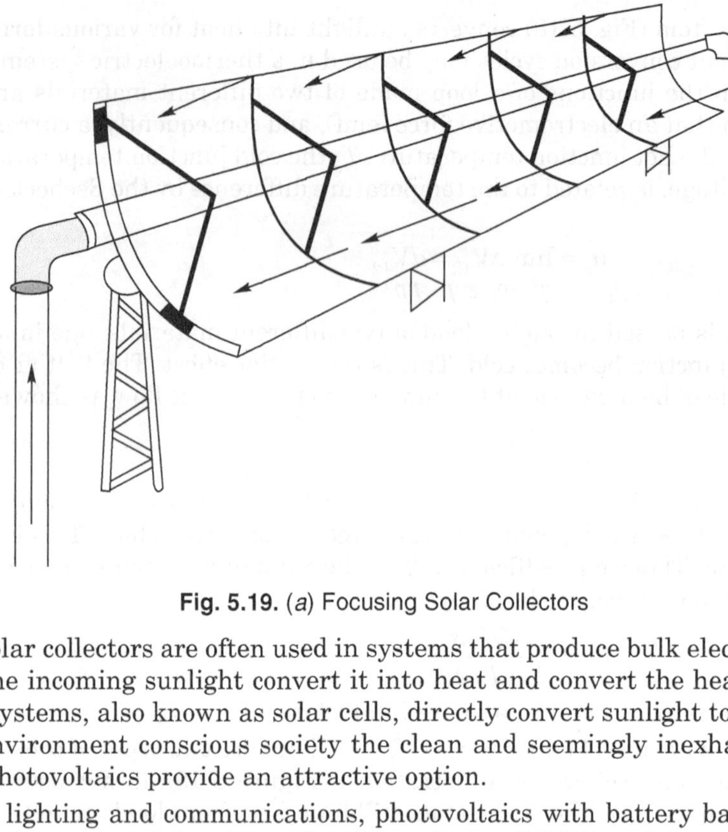

Fig. 5.19. (*a*) Focusing Solar Collectors

Focusing solar collectors are often used in systems that produce bulk electricity. They first concentrate the incoming sunlight convert it into heat and convert the heat into electricity. Photovoltaic systems, also known as solar cells, directly convert sunlight to electricity. In an energy and environment conscious society the clean and seemingly inexhaustible source of energy from photovoltaics provide an attractive option.

For remote lighting and communications, photovoltaics with battery backup provide the most cost-effective source of electricity. The photovoltaic cell (PV), module and array are shown in Figure 5.19 (*b*).

PV Devices

The solar cell is the basic building block of a PV system. It converts solar energy directly into electricity through a process known as the photovoltaic effect.

An individual solar cell doesn't generate much power. The amount varies, but it is around 1 or 2 watts. But you can increase the solar power generated by joining them together into a larger unit called a PV module. PV modules are also known as solar power panels. The power output of a PV module ranges from 10 watts to 300 watts. To generate even more power, PV modules can be combined to form a PV array.

A PV array can consist of any number of modules. This modularity makes photovoltaic systems very flexible. They are able to meet any electrical requirement, whether the amount of power needed is small or large.

Fig. 5.19. (*b*) PV Cell, Module, and Array

But to take advantage of the electricity produced by a PV array, you need a few more components.

Types of PV Systems

The three most common types of photovoltaic systems for residential solar power are:

- Grid-Tied PV System
- Grid-Tied PV System with Battery Backup
- Stand-Alone PV System.

Grid Tied PV System

A grid-tied PV system allows you to use the electricity generated by your PV system as well as electricity from the grid. When your PV system is producing electricity, your home will be powered by solar electricity. During the times when your PV system isn't producing electricity, such as at night, your home will receive power from the grid.

Any excess electricity produced by your system can be fed back to the grid. This is known as net-metering. With net-metering, when you use electricity from the grid your meter spins forward and when you are providing electricity to the grid your meter spins backwards. This offset means you are actually receiving market rates for the electricity you provide to the grid!

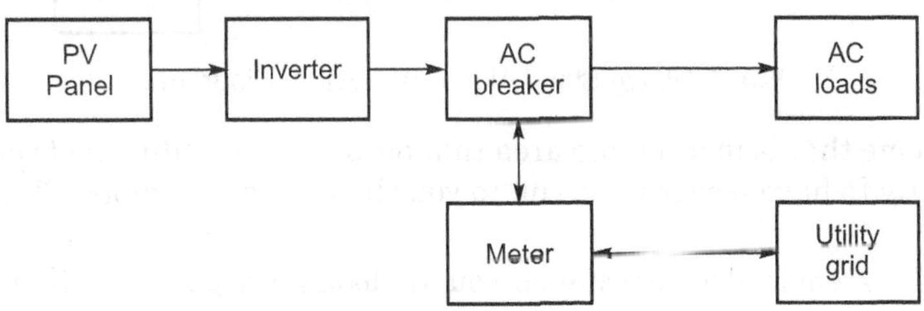

Fig. 5.19. (*c*) Grid Tied PV System Diagram

For most people, a grid-tied PV system is the ideal configuration. You get all the benefits of using your own solar electricity as well as the benefits of being connected to the grid. It is also less expensive than other types of PV systems.

Grid Tied PV System with Battery Backup

A grid-tied PV system can also be supplemented with a battery backup.

Why would you need a battery backup since you are connected to the grid?

Well, grid-tied PV system inverters are designed to shut down when the grid experiences a power outage in order to protect the utility repair workers from being shocked by electricity coming from your PV array. As a result, during a power outage, you won't be able to utilize the electricity coming from your PV array.

However, if your PV system includes a battery bank, during a power outage you can utilize the energy that your PV system produced and stored in the batteries.

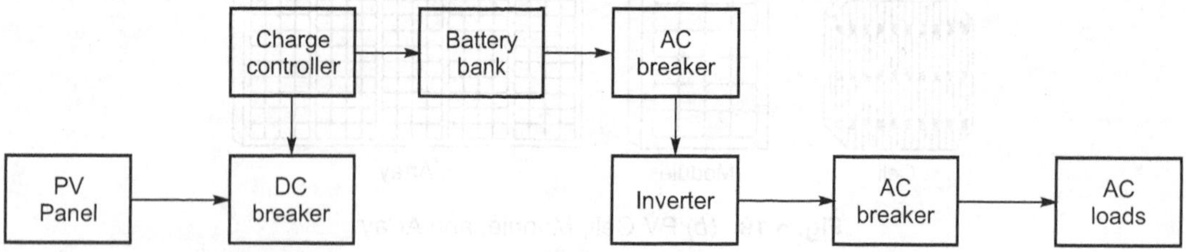

Fig. 5.19. (*d*) Grid Tied PV System with Battery Backup Diagram

A grid-tied PV system with battery backup is ideal if you live in an area that has unreliable power from the grid or that experiences power outages due to natural disasters.

Stand Alone PV System

Stand-alone PV systems are designed to operate independently from the grid and to provide all of the electricity you need for your home.

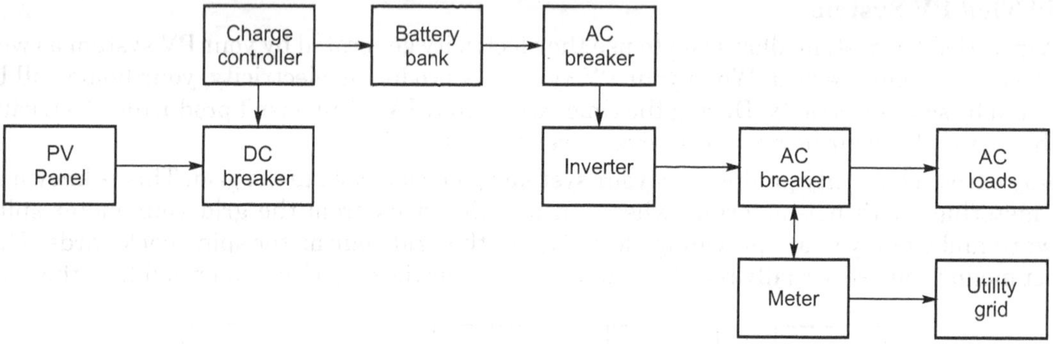

Fig. 5.19. (*e*) Stand Alone PV System Diagram

If you have a home that is in a remote area that doesn't have utility grid service or where it would be very costly to have power lines run to your home, a stand-alone PV system would be ideal.

This type of PV system is also desirable if you are looking to go completely green with your source of electricity.

Photovoltaic Effect

The photovoltaic effect (Figure 5.20) is most prominent in various semiconductors. Most commercial solar cells are made of crystalline and amorphous silicon materials. When sunlight strikes the solar cell, part of the tight spectrum imparts enough energy to create electron-hole pairs in the semiconductor material. A potential barrier in the cell is set up by forming a junction between dissimilarly doped semiconductor layers.

This separates the light-generated carriers (*i.e.*, electrons and holes) resulting in an induced voltage of about 1/2 volt. The available current is a function of cell area and light intensity. The electricity is collected and transported by metallic contacts placed on both surfaces of the cell.

For even greater power modules can be interconnected in larger groups to form arrays. The *dc* electricity generated by the solar cell array is usually passed through a power conditioner for voltage and power regulation and conversion to alternating current.

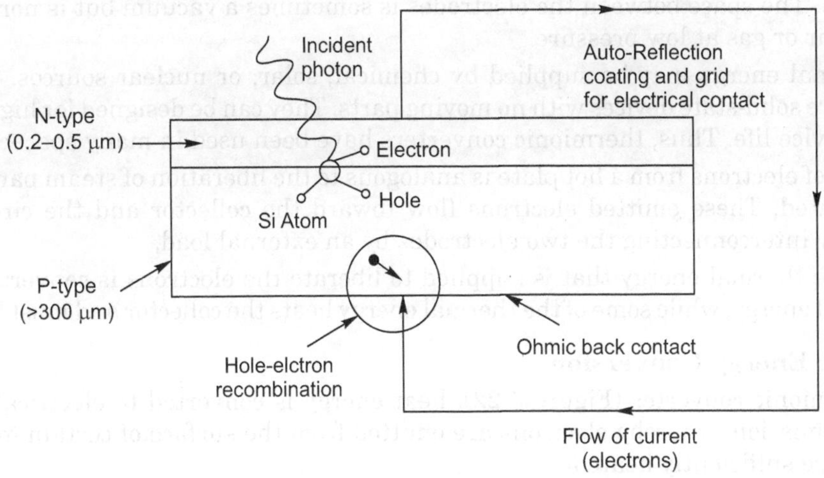

Fig. 5.20. Photovoltaic Effect

Commercially available photovoltaic modules can convert sunlight into electricity with efficiencies ranging from 5% to 15%. The cost of photovoltaic cells has dropped from ₹ 50,000 per peak watt in the 1950's to under ₹ 250 per peak watt. So that they can produce electricity for as little as 25 to 30 cents per kilowatt-hour.

5.10 THERMIONIC POWER GENERATION

Any class of devices that convert heat energy directly into electricity using thermionic emission rather than first changing it to some other form of energy is called thermionic power converters (Figure 5.21).

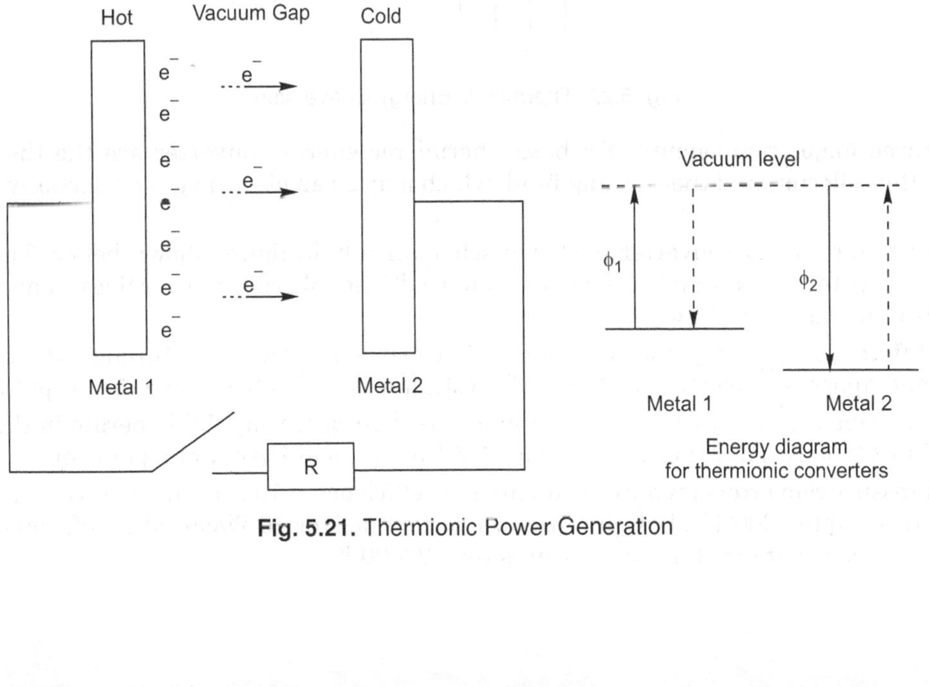

Fig. 5.21. Thermionic Power Generation

A thermionic power converter has two electrodes. One of these is raised to a sufficiently high temperature to become a thermionic electron emitter or "hot plate". The other electrode, called a collector because it receives the emitted electrons, is operated at a significantly lower temperature. The space between the electrodes is sometimes a vacuum but is normally filled with a vapour or gas at low pressure.

The thermal energy may be supplied by chemical, solar, or nuclear sources. Thermionic converters are solid-state devices with no moving parts. They can be designed for high reliability and long service life. Thus, thermionic converters have been used in many spacecraft.

Emission of electrons from a hot plate is analogous to the liberation of steam particles when water is heated. These emitted electrons flow toward the collector and the circuit can be completed by interconnecting the two electrodes by an external load.

Part of the thermal energy that is supplied to liberate the electrons is converted, directly into electrical energy, while some of the thermal energy heats the collector and must be removed.

Thermionic Energy Conversion

In a thermionic converter (Figure 5.22), heat energy is converted to electrical energy by thermionic emission, whereby electrons are emitted from the surface of certain metals when the metals are sufficiently heated.

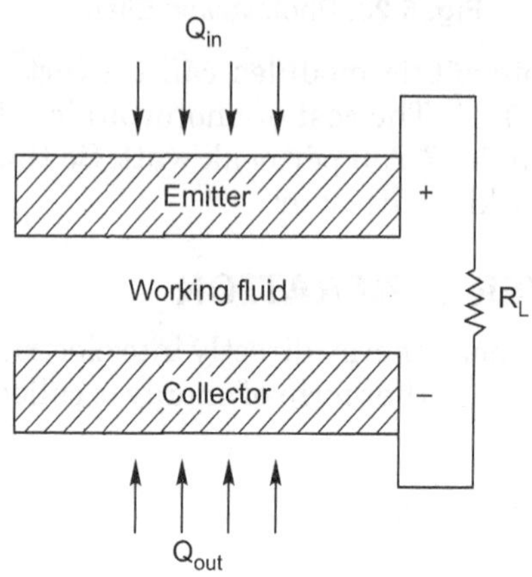

Fig. 5.22. Thermionic Energy Conversion

The three major components of a basic thermionic energy converter are the thermionic emitter, the collector, and the working fluid, which may be an electron gas or partially ionized plasma.

A thermionic energy converter is shown schematically in figure shown below. The input heat Q_{in}' heats the emitter and electrons are emitted. The cold collector receives some of these electrons at an output heat, Q_{out}.

The difference (Q_{in} – Q_{out}) is the energy that drives the electrons through the external circuit, and appears as electrical energy. The collector is cooled to remove the output heat.

Vacuum converters, which have electron gas as their operating fluid operate in the range 12000 K to 14000 K. They typically produce 1 W/cm^2 at an efficiency of 5 per cent.

Low-pressure converters produce 10 W/cm^2 at an efficiency of 10 per cent operating at emitter temperatures upto 2300°K. High-pressure converters deliver 40 W/cm^2 at an efficiency of 20 per cent, operating at emitter temperatures upto 22000 K.

Principal applications of thermionic energy converters are in regions not easily accessible, such as outer space, undersea and Polar Regions. The two important heat sources are the sun and nuclear reactors.

5.11 SOLAR INVERTER

5.11.1 How an Inverter Works

So how can an inverter give us a high voltage alternating current from a low voltage direct current.

Let's first consider how an alternator produces an alternating current. In its simplest form, an alternator would have a coil of wire with a rotating magnet close to it. As one pole of the magnet approaches the coil, a current will be produced in the coil. This current will grow to a maximum as the magnet passes close to the coil, dying down as the magnetic pole moves further away. However when the opposite pole of the magnet approaches the coil, the current induced in the coil will flow in the opposite direction.

As this process is repeated by the continual rotation of the magnet, an alternating current is produced.

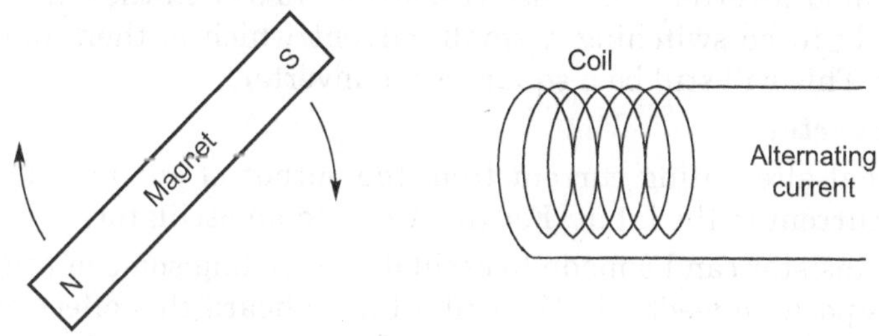

Fig. 5.23

Now lets consider what a transformer does. A transformer also causes an electric current to be induced in a coil, but this time, the changing magnetic field is produced by another coil having an alternating current flowing through it. Any coil with an electric current flowing through it will act like a magnet and produce a magnetic field. If the direction of the current changes then the polarity of the field changes.

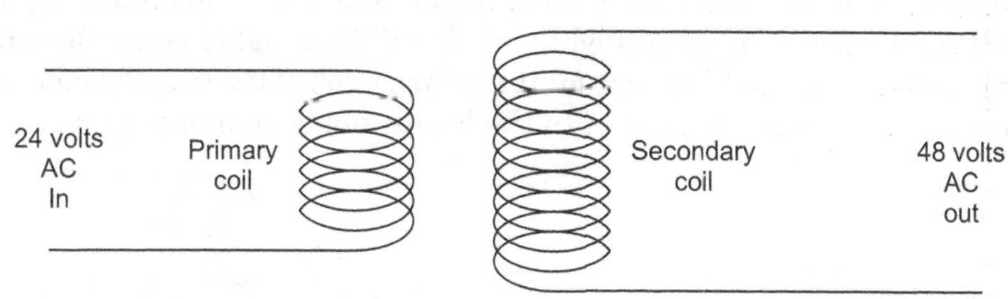

Fig. 5.24

Now, the handy thing about a transformer is that, the voltage produced in the secondary coil is not necessarily the same as that applied to the primary coil. If the secondary coil is twice the size (has twice the number of turns) of the primary coil, the secondary voltage will be twice that of the voltage applied to the primary coil. We can effectively produce whatever voltage we want by varying the size of the coils.

If we connected a direct current from a battery to the primary coil it would not induce a current in the secondary as the magnetic field would not be changing. However, if we can make that direct current effectively change direction repeatedly, then we have a very basic inverter. This inverter would produce a square wave output as the current would be changing direction suddenly.

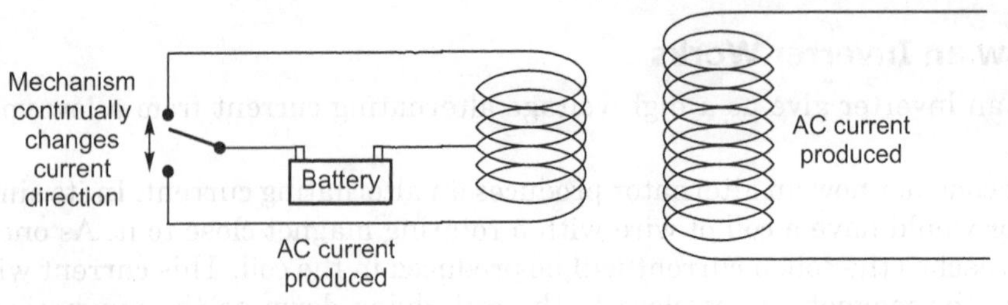

Fig. 5.25

This type of inverter might have been used in early car radios that needed to take 12 volts available in the car and produce the higher voltages required to run radio valves (known as tubes in America) in the days before transistors were widely used.

A more sophisticated inverter would use transistors to switch the current. The switching transistors are likely to be switching a small current which is then amplified by further transistor circuitry. This will still be a square wave inverter.

The Sine Wave Inverter

To get a sinusoidal alternating current from the output of our transformer, we have to apply a sinusoidal current to the input. For this we need an oscillator.

An amplifying transistor can be made to oscillate by feeding some of the amplified output back to its input as positive feedback. We will all have heard this effect at sometime when someone is setting up a PA or microphone system. If the microphone is too close to the speaker, some of the output from the speaker is fed back to the microphone and inputted to the amplifier again. The result is a howling sound.

The positive feedback in an electronic circuit can be tuned using extra components to produce the frequency we require (generally either 50 or 60 cycles per second to mimic mains electricity). If a crystal is used to control this frequency, as in a battery watch or clock, the frequency can be very accurately controlled.

As with simpler switching transistor circuit, the oscillator will be producing a low current output. This will then need to be amplified by what will be roughly equivalent to a powerful audio amplifier to produce the high current for the primary coil of the transformer (the frequency of mains AC current is roughly equivalent to the lowest notes on a bass guitar).

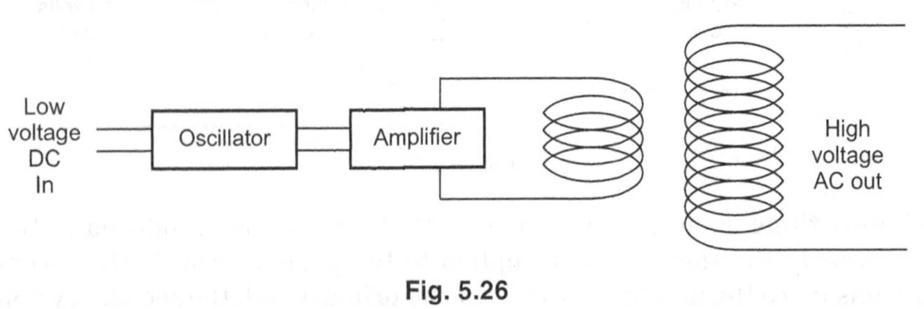

Fig. 5.26

The transformer, while being very useful, does not do something for nothing. While increasing the voltage, the current will be reduced, and the power (voltage × current) will stay the same (less any inefficiency of the transformer). In other words, to get 1 kW of high voltage AC current out, you have put 1 kW of low voltage AC current in.

Grid Tied Inverters

If the above example were a grid tied inverter, *i.e.* able to feed power back into the national grid, it would need to use a sample of the mains voltage to then be amplified within the inverter, or to synchronise the oscillator with that sample.

Grid tied inverters will also sense if there is a "power cut" and disconnect themselves from the grid. If they did not have this facility, in the event of a power cut, your inverter would be attempting to power all your neighbours houses and would present an electrocution risk to anyone working on power lines that had supposedly been turned off.

Inverter Stacking: Using multiple inverters

Two inverters can be installed in a configuration known as stacking that can provide more power or higher voltage. If two compatible inverters are stacked in *series* you can double the output voltage. This would be the technique to use to provide 120/240 volts AC. On the other hand, if you configure them in *parallel,* you can double your power. Two 4000 watt inverters in parallel would give you 8000 watts (8 kW) of electricity.

What is a Solar Inverter?

For more information about the basics of electricity, take a look at Section 5.11.

Solar Power Electricity

The purpose of an **inverter** is to change Direct Current (DC) electricity into Alternating Current (AC) electricity. It also will increase the voltage of the AC electricity to 120 volts AC (if you're in the United States) or 240 volts AC (if you're in most other parts of the world).

But why is a solar power inverter needed in the first place?

Well, a **solar inverter** is needed because the electricity generated by your solar panels is DC electricity. In order to use the generated solar power with your electrical devices and household appliances, it needs to be converted to the standard voltage AC electricity for your region.

A **solar inverter** or **PV inverter** is a critical component in a photovoltaic system. It converts the variable DC output of the solar panel into a utility frequency alternating current that can be fed into the commercial electrical grid or used by a local, off-grid electrical network. An inverter allows use of ordinary mains-operated appliances on a direct current system. Solar inverters have special functions adapted for use with PV arrays, including *maximum power point tracking* and *anti-istanding protection.*

Types of Solar Inverters

Solar inverters can be categorized by the type of solar power system they are used with

- Grid Tie Inverter
- Off Grid Inverter
- On/Off Grid Inverter

A Grid Tie Inverter is used with a *grid-tied PV system.* This type of solar power inverter takes the DC electricity generated by the solar panels and converts it to AC electricity directly. It channels the AC electricity to your home's breaker panel where it is either used by your household loads or, if you are signed up for net-metering, sends it to the utility grid.

An **Off Grid Inverter** is used with a *stand-alone* or *off-grid solar system.* An off-grid solar system typically uses batteries to store the DC electricity generated by your solar panels. The inverter converts the DC electricity from the batteries into AC electricity for use in your home.

An **On/Off Grid Inverter** is used with a *grid-tied PV system with battery backup.* This type of inverter allows you to connect your home to the utility grid and use power from a battery bank. Unlike a pure grid tie inverter, this type of inverter is able to continue transferring power to your home when the utility has a blackout. During a blackout, a grid tie inverter will not be operational.

5.12 LIGHTING ARRESTER

NEPSI's Metal-Enclosed Capacitor Banks and Harmonic Filter Banks come equipped with heavy duty distribution class polymer lightning arresters. These arresters are typically connected to the load-side terminals of the incoming air-disconnect switch. If desired, they can be connected to the line side terminals.

As an option, to provide a higher level of protection, the arresters can be changed to either station class or intermediate class. The lightning arresters are typically manufacturing red by either Cooper Power Systems or General Electric. Arresters from other manufacturers can be supplied upon request.

Arrester voltage rating is chosen based on system grounding and expected normal and abnormal system over-voltage.

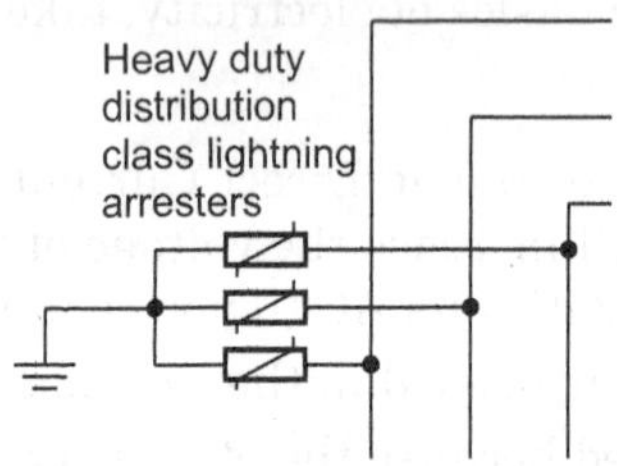

Fig. 5.27

Lightning Arresters help reduce the possibility of damage from lightning and switching transients.

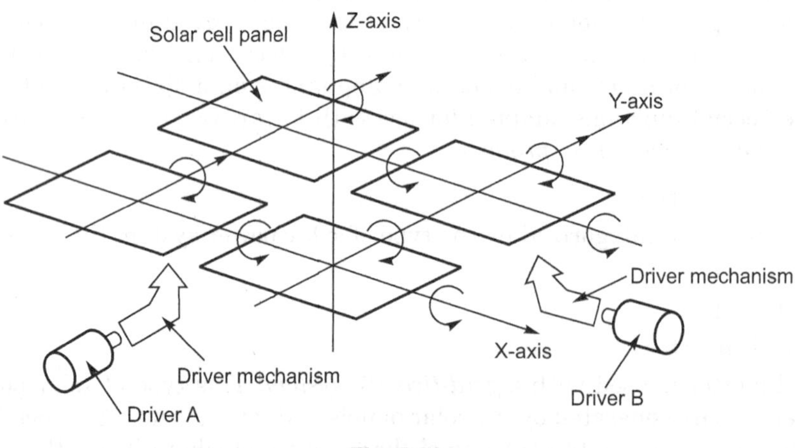

Fig. 5.28

Most photovoltaic (PV) solar panels are fitted in a fixed location—for example, on the sloping roof of a house, or on framework fixed to the ground. Since the sun moves across the sky through the day, this is far from an ideal solution.

Solar panels are usually set up to be in full direct sunshine at the middle of the day facing South in the Northern Hemisphere, or North in the Southern Hemisphere. Therefore morning and evening sunlight hits the panels at an acute angle reducing the total amount of **electricity** which can be generated each day.

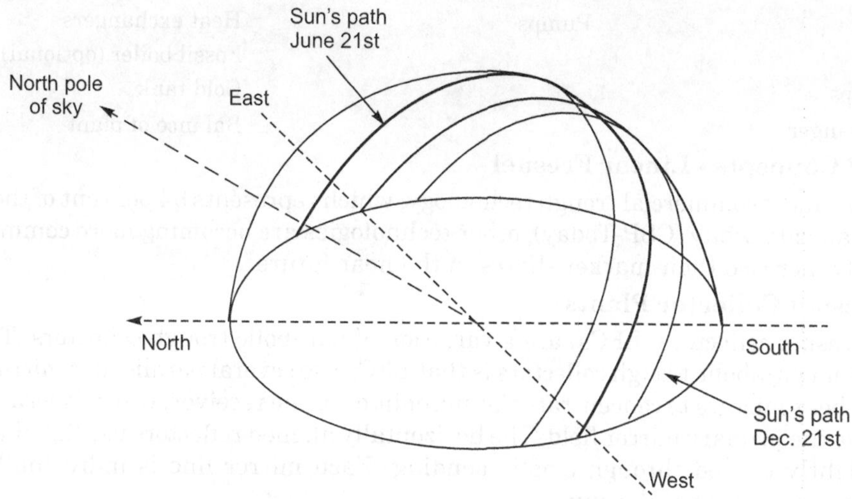

Fig. 5.29

During the day the sun appears to move across the sky from **left to right** and **up and down** above the horizon from sunrise to noon to sunset. This is shown in the schematic above of the Sun's apparent motion as seen from the Northern Hemisphere.

Parabolic Trough Power Plants

The main components of parabolic trough power plants of the single components can be to provide the basis for the subsequent analyses of the manufacturing processes, of the cost of components and processes, and of the potential to produce components in MENA countries. The analysis of the components is based on state of the art technology, which consists of a parabolic trough using thermal oil as heat-transfer fluid and the power block. Optionally, a thermal energy storage can be used:

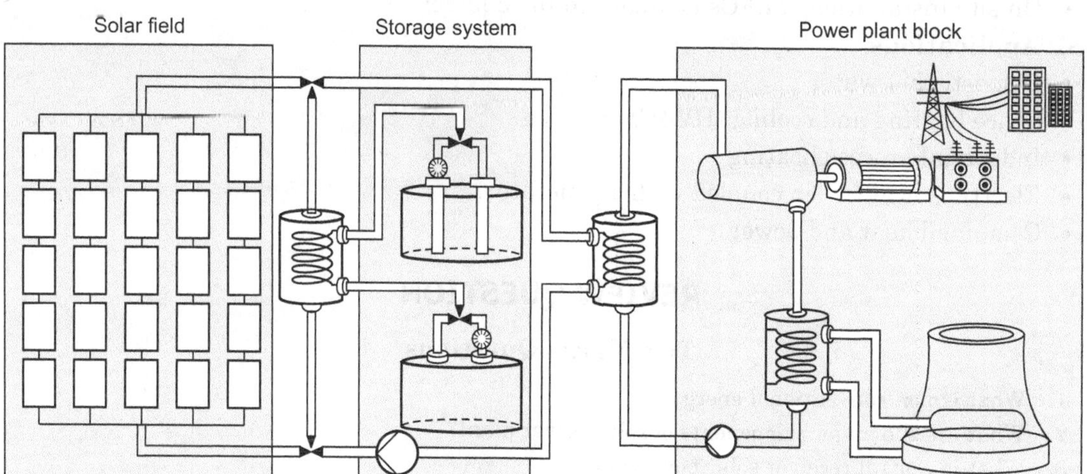

Fig. 5.30. Parabolic Trough Power Plant

Solar field	Thermal storage	Power block
Receiver	Molten salt	Turbine
Mirror	Hot tank	Generato
Support structure	Cold tank	Condenser
Tracking	Heat exchangers	Pumps
Piping	Pumps	Heat exchangers
HTF (oil)		Fossil boiler (optional)
HTF pumps		Cold tank
Heat exchanger		Balance of plant

Other CSP Concepts - Linear Fresnel

Beyond the most commercial trough technology, which represents 94 per cent of the installed CSP plant capacity today (CSP-Today), other technologies are becoming more commercial and will probably increase their market shares in the near future.

Linear Fresnel Collector Plants

Linear Fresnel collectors (LFCs) are a variation of parabolic trough collectors. Their main difference from parabolic trough collectors is that LFCs use several parallel flat mirrors instead of parabolic bent mirrors to concentrate the sunlight onto one receiver, which is located several meters above the primary mirror field. The horizontally aligned reflectors use flat glass mirrors that are slightly curved through elastic bending. Each mirror line is individually tracked according to the position of the sun.

The receiver also consists of a long, selectively coated absorber tube, without any need for the flexible hoses or rotating connectors required by a parabolic trough. Due to the optical principles of Fresnel collectors, the focal line is distorted by astigmatism (Mertins, 2009). This requires a secondary mirror above the tube to refocus the rays missing the tube in a secondary reflection onto the tube. Another concept is based on several parallel tubes forming a multi-tube receiver, thereby increasing the width instead of using a secondary reflector.

The main differences between the Fresnel concept and the parabolic trough collector include:

- LFCs use cheap, flat mirrors (6–20 C/m^2) instead of expensive parabolic curved mirrors (25–30 C/m^2); furthermore, flat glass mirrors are a standardized mass product.
- LFCs require less heavy steel material, using a metal support structure with limited or no concrete (making for easier assembly).
- On-site installation of LFCs is predicted to be faster.

CSP Applications

- Domestic hot water
- Space heating and cooling (HVAC)
- Industrial process heating
- Electric power (when coupled with a generator)
- Combined heat and power

REVIEW QUESTION

Two Marks Questions

1. What is non-conventional energy?
2. What are the major components of solar power plant?
3. What is central receiver solar power plant?
4. What is concentration ratio?

5. What is solar chimney system?
6. What is solar energy storage?
7. What is space heating and space cooling?
8. What are the solar energy collectors?
9. What are the applications of solar collectors?
10. What are the solar thermal applications?
11. What is solar photovoltaic?
12. What is wind power generator?
13. What is tidal power generator?
14. What are the factors to be considered for suitable site selection of tidal power plant?
15. Mention the advantages and disadvantages of geothermal power plant.
16. What are the fixed costs?
17. List out the advantages of tidal plant over hydel power plant.
18. What are the major factors that decide the economy of the power plant?
19. What are the advantages and limitations of tidal power generation?
20. What do you understand by the term tariff?
21. What are the components of tidal power plants?
22. How does the fuel cost relate to the load and the cost of power generation?
23. What are the important criteria while selecting the geothermal energy?
24. What is the basic principle of tidal power?
25. What is a fuel cell? Give its types.
26. What is meant by reversible fuel cell?
27. What of heliostat? Give its basic aspects.
28. What is geothermal power plant?
29. What are the main parts of geothermal power plant?
30. What are the type geothermal energy conversion systems?
31. How the OTEC is classified?
32. What are the benefits of OTEC?
33. What are the disadvantages of OTEC?
34. What is meant by hybrid?
35. What is geothermal energy? Give its applications.
36. Give some geothermal energy sources in India.
37. What are the different geothermal fluids and give its temperature range?
38. What is biogas generating?
39. What are the types of bioconversion process?
40. What are the raw materials used for producing biogas?
41. What are the properties of biogas?
42. What are the types of biogas plants?
43. What are the salient features of biogas plant?
44. What is fuel cell?
45. What are the types of fuel cell?
46. What are the classifications of MHD power generator?
47. What are the methods of ionizing the gas?
48. Define the following:
    (*i*) Faraday current
    (*ii*) Plasma
    (*iii*) Hall effect current
    (*iv*) Efficiency.

49. What is thermo electric power generating system?

50. Define photovoltaic effect.

51. What is thermionic power generation and thermionic energy conversion?

Sixteen Marks Questions

1. Explain the working principle of solar power plant with suitable sketch.

2. Explain the working principle of wind power generation with neat sketch.

3. Explain the working principle of tidal power generation with near sketch.

4. Explain with neat sketch the working principle of geothermal power plant.

5. What are the types of geothermal power generator? Explain with sketch.

6. Explain the working principle of OTEC with a neat diagram.

7. Explain the working principle of biogas generation, with suitable sketch.

8. What are the types of biogas plant? Explain.

9. Explain the working principle of floating gas holder type biogas plant with neat sketch.

10. Explain the working principle of fuel cell with suitable sketch.

11. Explain the working principle of the MHD power plant with a heat sketch.

12. Explain the working principle of thermoelectric power generation system.

13. Explain solar thermal system.

14. Explain the following with a neat sketch:

(*i*) Photovoltaic effect

(*ii*) Thermionic power generator

(*iii*) Thermionic energy conversion.

University Question Papers

B.E./B.Tech DEGREE EXAMINATION, NOVEMBER/DECEMBER 2011

Fourth Semester

Electrical and Electronics Engineering

EE 2252—POWER PLANT ENGINEERING

(Regulation 2008)

Time: Three hours Maximum : 100 Marks

Answer ALL Questions

Part A—(10 × 2 = 20 marks)

1. What is the necessity of feed pump in thermal power plant?
2. Mention the advantages of pulverized fuel firing system.
3. What is the function of draft tube in hydroelectric power plant?
4. What are the different types of spill ways?
5. What is known as binding energy?
6. What is the function of pressurizer in pressurized water reactor?
7. List a few differences between the closed cycle and open cycle gas turbine power plant.
8. What are the functions of lubrication system?
9. What are the factors to be considered for suitable site selection of tidal power plant?
10. What are the different sources of geothermal energy?

Part B—(5 × 16 = 80 marks)

11. (*a*) (*i*) Explain with the schematic sketch he working of different circuits of the thermal power plant. (10)

(*ii*) Name the various equipment used for transferring the coal. Explain the working principle of any one equipment. (6)

Or

(*b*) (*i*) Draw a neat diagram of cyclone burner and describe it's working. What are its outstanding features compared with other burners? (8)

(*ii*) Explain with neat sketch the fluidized bed combustion system. (8)

12. (*a*) Explain with a neat sketch the essential elements of a hydroelectric power plant. (16)

Or

(*b*) (*i*) Explain with a neat sketch the function of surge tank in hydroelectric power plant. (8)

(*ii*) How are dams classified? What are the factors to be considered in selecting a type of dams? (8)

13. (*a*) (*i*) With a neat sketch explain the functions of different components of nuclear reactor. (8)

(*ii*) What is known as nuclear fission? What are the essential requirements to cause nuclear fission? (8)

Or

(*b*) (*i*) With a neat sketch explain the working pressurized water reactor in nuclear power plant. (8)

(*ii*) Explain with a neat sketch the difference between controlled and uncontrolled nuclear chain reaction. (8)

14. (*a*) (*i*) Discuss the effect of intercooling and reheating in a gas turbine power plant. (8)

(*ii*) Explain with a neat sketch the semi-closed cycle gas turbine power plant. (8)

Or

(*b*) (*i*) List the various components of diesel engine power plant and explain its functions with neat sketch. (10)

(*ii*) Give the advantages and disadvantages of diesel engine power plants. (6)

15. (*a*) (*i*) With a neat sketch explain the working principle of fuel cell and give its advantages. (8)

(*ii*) Explain the working principle of a vertical axis wind mill. (8)

Or

(*b*) (*i*) Explain the principle of working of magneto hydro dynamic system. (8)

(*ii*) Give the classification of tidal power plants and explain its working principle of any one type. (8)

B.E./B.Tech DEGREE EXAMINATION, APRIL/MAY 2010

Fourth Semester

Electrical and Electronics Engineering

EE 2252—POWER PLANT ENGINEERING

(Regulation 2008)

Time: Three hours Maximum : 100 Marks

Answer ALL Questions

Part A—(10 × 2 = 20 marks)

1. Mention the various modern ash handling systems.
2. What are the factors affecting cooling of water in cooling tower?
3. Enumerate advantages and disadvantages of hydropower plants.
4. What is a surge tank?
5. What are the requirements of fission process?
6. What are the essential components of a nuclear reactor?
7. State the merits and demerits of closed cycle gas turbine over open cycle gas turbine power plant.
8. State the applications of diesel power plant.
9. What is a fuel cell?
10. What are the advantages and limitations of tidal power plant?

Part B—(5 × 16 = 80 marks)

11. (*a*) Give the layout of a modern steam power plant and explain it briefly. (16)

Or

(*b*) How are cooling towers classified? Explain any one of them with a neat sketch. (16)

12. (*a*) Explain with a neat sketch a pumped storage plant. (16)

Or

(*b*) What is the function of a hydraulic turbine? How are the turbines classified? Explain anyone with a suitable sketch. (16)

13. (*a*) With the help of a sketch show all the important part of nuclear reactor. Describe briefly the functions of each part. (16)

Or

(*b*) Explain the following terms: (*i*) Fission of nuclear fuel (*ii*) Distribution of fission energy (*iii*) The chain reaction. (16)

14. (*a*) Discuss briefly the methods employed for improvement of thermal efficiency of open gas turbine power plant. (16)

Or

(*b*) List the essential components of a diesel power plant and explain them briefly. (16)

15. (*a*) Explain with a neat diagram of wind electric generating power plant. (16)

Or

(*b*) Describe open cycle MHD system with suitable sketch. (16)

B.E./B.Tech DEGREE EXAMINATION, NOVEMBER/DECEMBER 2010

Fourth Semester

Electrical and Electronics Engineering

EE 2252—POWER PLANT ENGINEERING

(Regulation 2008)

Time: Three hours Maximum : 100 Marks

Answer ALL Questions

Part A—(10 × 2 = 20 marks)

1. What are the processes in a Rankine cycle?
2. Define FBC?
3. What is the function of a surge tank in the hydroelectric power plant?
4. What is the purpose of trash rack?
5. What are the requirements of fission process?
6. What is the function of a moderator?
7. What are the applications of gas turbine plant?
8. List out the various components in a diesel engine power plant?
9. Define Photovoltaic cell?
10. Define MHD power generation?

Part B—(5 × 16 = 80 marks)

11. (*a*) (*i*) Draw a general layout of thermal power plant and explain the working of different circuits. (10)

(*ii*) Describe the operation of a balanced draught system with neat sketch. (6)

Or

(*b*) (*i*) Name the various methods of ash handling. Describe the pneumatic system of ash handling. (8)

(*ii*) Explain the various steps involved in "in plant coal handling in a thermal plant". (8)

12. (*a*) (*i*) What are the factors to be considered while selecting a site for hydroelectric power plant? (8)

(*ii*) Draw the layout of a hydroelectric power plant and explain its functions of components? (8)

Or

(*b*) (*i*) Describe the selection factors for a hydraulic turbines? (8)

(*ii*) Explain with a neat sketch the principle of pumped storage plant? (8)

13. (*a*) (*i*) Explain with neat sketch the working of fast breeder reactor plant? (8)

(*ii*) What are the advantages and disadvantages of nuclear power plant? (8)

Or

(*b*) What is chain reaction? How is it controlled? Explain with the neat sketch a boiling water reactor? (16)

14. (*a*) State the advantages and disadvantages of gas turbine power plants over diesel and thermal power plants? (16)

(*b*) Draw the layout of diesel engine power plant and explain the functions of various components of diesel engine power plant? (16)

15. (*a*) Discuss the three types of direct energy conversion devices? (16)

Or

(*b*) Write short notes on the following plants:

(*i*) Wind power plant (8)

(*ii*) Tidal power plant. (8)

B.E./B.Tech DEGREE EXAMINATION, MAY/JUNE 2013

Fourth Semester

Electrical and Electronics Engineering

EE 2252—POWER PLANT ENGINEERING

(Regulation 2008/2010)

Time: Three hours Maximum : 100 Marks

Answer ALL Questions

Part A—(10 × 2 = 20 marks)

1. What is the function of not primary air?
2. What is super-critical boiler? Give any two advantages.
3. What is the function of draft tube?
4. List any four advantages of hydroelectric power plant.
5. What is an LMFBR? Why is a liquid metal the preferred coolant in a fast breeder reactor?
6. What do you mean by mass defect?
7. How the solid injection system is classified?
8. What do you mean by regeneration in gas turbine power plant?
9. What do you understand by zero energy houses?
10. What are the classifications of geothermal energy?

Part B—(5 × 16 = 80 marks)

11. (*a*) Describe the different types of overfeed stokers and discuss the merits and demerits of each over others. (16)

Or

(*b*) (*i*) Explain with a neat sketch the function of Benson Boiler and give its advantages. (8)

(*ii*) Briefly explain the air-cooled cooling system. (8)

12. (*a*) (*i*) Enlist the advantages and disadvantages of water power. (8)

(*ii*) What is spill ways? Briefly explain different types of spillways with sketch. (8)

Or

(*b*) (*i*) How does a pumped hydro system operate? Show the main components in a neat sketch of the system. (10)

(*ii*) Discuss the criteria to be considered while selecting site for dam construction. (6)

13. (*a*) (*i*) Explain different types of nuclear reactions and initiation of nuclear reactions. (8)

(*ii*) Briefly explain the pressurized water reactor (PWR) with neat sketch. (8)

Or

(*b*) (*i*) Explain Boiling Water Reactor (BWR) with neat sketch. Give its advantage and disadvantage. (8)

(*ii*) Explain different methods for nuclear waste disposal with necessary sketch. (8)

14. (*a*) (*i*) Explain different components of gas turbine plant with neat sketch. (8)

(*ii*) Discuss the effect of intercooling and reheating in a gas turbine plant. (8)

Or

(*b*) (*i*) Explain water cooling system in diesel power plants with neat sketch. (10)

(*ii*) What is an engine day tank? State the functions of a fuel injection system. (6)

15. (*a*) (*i*) Briefly explain the classification of tidal power plant with neat sketch. (8)

(*ii*) Briefly explain the low temperature system with flat plate collector in solar power plants. (8)

Or

(*b*) (*i*) Explain different types of MHD generators with neat sketch. (10)

(*ii*) Briefly explain the working principal of fuel cell. (6)

Index